DFG

**The MAK-Collection for
Occupational Health and Safety**

Part IV: Biomonitoring Methods

The MAK-Collection

Current Volumes

Part I: MAK Value Documentations, Volume 24
Greim, H. (ed.)

2007. ISBN 978-3-527-31594-9

Part II: BAT Value Documentations, Volume 4
Drexler, H. and Greim, H. (eds.)

2005. ISBN 978-3-527-27049-1

Part III: Air Monitoring Methods, Volume 10
Parlar, H. and Greim, H. (eds.)

2007. ISBN 978-3-527-31601-4

The MAK-Collection online
www.mak-collection.com

DFG Deutsche Forschungsgemeinschaft

The MAK-Collection for Occupational Health and Safety

Part IV: Biomonitoring Methods

Volume 11

Edited by Jürgen Angerer
Working Group Analytical Chemistry

Commission for the Investigation of Health Hazards of Chemical Compounds in the Work Area
(Chairwoman: Andrea Hartwig)

WILEY-VCH Verlag GmbH & Co. KGaA

Prof.-Dr. Andrea Hartwig
DFG-Senatskommission zur Prüfung
gesundheitsschädlicher Arbeitsstoffe
Gustav-Meyer-Allee 25
13355 Berlin
Germany

Prof. Dr. Jürgen Angerer
Institut für Arbeits- und
Sozialmedizin
Universität Erlangen-Nürnberg
Schillerstr. 25/29
91054 Erlangen
Germany

Vol. 1–9 were published under
the title "Analyses of Hazardous
Substances in Biological
Materials"
(ISSN 0179-7247)

1st Edition 2008
 1st Reprint 2008

■ All books published by Wiley-VCH
are carefully produced. Nevertheless,
authors, editors, and publisher do not
warrant the information contained in
these books, including this book, to be
free of errors. Readers are advised to
keep in mind that statements, data,
illustrations, procedural details or other
items may inadvertently be inaccurate.

Library of Congress Card No.: applied for

British Library Cataloging-in-Publication Data: A catalogue record for this book is available from the British Library.

Bibliographic information published by Die Deutsche Bibliothek
Die Deutsche Bibliothek lists this publication in the Deutsche Nationalbibliografie; detailed bibliographic data is available in the Internet at http://dnb.ddb.de.

© 2008 WILEY-VCH Verlag GmbH & Co. KGaA, Weinheim

All rights reserved (including those of translation into other languages). No part of this book may be reproduced in any form – by photoprinting, microfilm, or any other means – nor transmitted or translated into a machine language without written permission from the publishers. Registered names, trademarks, etc. used in this book, even when not specifically marked as such, are not to be considered unprotected by law.

Typesetting K+V Fotosatz GmbH, Beerfelden
Printing betz-druck GmbH, Darmstadt
Binding Litges & Dopf GmbH, Heppenheim

Printed in the Federal Republic of Germany
Printed on acid-free paper

ISSN: 1860-4994
ISBN: 978-3-527-31596-3

Preface

At a very early stage the Deutsche Forschungsgemeinschaft's Commission for the Investigation of Health Hazards of Chemical Compounds in the Work Area (MAK Commission) recognised the possibilities offered by determination of chemical substances and their metabolites in human body fluids for the prevention of health impairment. Therefore the DFG established an "Analyses in Biological Material" working group in the early 1970s. The task of this group was to devise reliable analytical procedures as a basis on which the performance of human biomonitoring would be possible. To achieve this goal the analytical methods not only had to be validated but also described in such detail that they could be replicated in other laboratories. Thus all the methods contained in these volumes have been subjected to rigorous experimental testing in the laboratory. The concept underlying this work is that only valid and comparable results of human biomonitoring can offer the same health protection for everyone. The quality of the analytical methods has therefore always had priority over commercial considerations.

Today a collection of methods is available in German and English that enables the assay in body fluids of more than 250 of the most significant substances to occupational and environmental medicine and their metabolites. This collection of methods has been supplemented for the 11th time by eleven further analytical methods and by a monograph on the use of LC/MS in the field of human biomonitoring. On the basis of these analytical principles human biomonitoring has advanced to become an indispensable tool for occupational and environmental medicine.

As the chairwoman of the Deutsche Forschungsgemeinschaft's Commission for the Investigation of Health Hazards of Chemical Compounds in the Work Area I wish to thank all the present and past members and guests of the "Analyses in Biological Material" working group for their valuable contributions, which were provided in an honorary capacity. Moreover, I would like to extend special thanks to the Deutsche Forschungsgemeinschaft, which has continued to support this working group for many years, with the result that the German contribution to the field of human biomonitoring is regarded as excellent throughout the world.

Prof. Dr. Andrea Hartwig
Chairwoman of the Deutsche Forschungsgemeinschaft's
Commission for the Investigation of Health Hazards of
Chemical Compounds in the Work Area

Foreword

This is the 11th English edition of this collection of validated analytical methods developed and issued by the DFG's Commission for the Investigation of Health Hazards of Chemical Compounds in the Work Area. Once again this publication contains a general analytical chapter, in this case introducing the combination of liquid chromatography and mass spectrometry (LC/MS). Thereafter we present eleven methods for the assay of chemical substances or their metabolites in human fluids in detailed "Standard Operating Procedures" that, as always, have been checked for their reliability and reproducibility.

Although this work has gone only a little way to extending the analytical basis of human biomonitoring, the many supplements to this collection that have been published in the past 30 years have helped to elevate human biomonitoring to an established and accepted procedure for the prevention of health impairment.

Human biomonitoring has long proved its worth as part of preventive examinations in occupational medicine. It has even been made obligatory by law in Germany. Throughout the world human biomonitoring is being increasingly used to resolve questions that arise in the field of environmental medicine. It is widely applied, especially in Germany and in the USA, for representative determination and documentation of the exposure of the general population to chemical substances.

These chemical substances are common to investigations in both occupational and environmental medicine. Those that are widely handled at the workplace are generally also found in the environment. Therefore it has proved advantageous to devise analytical methods for human biomonitoring that are sufficiently sensitive to enable determination of the concentration ranges that are of relevance to both environmental medicine and occupational medicine. It is increasingly becoming unnecessary to separate the substances that are predominantly discussed from the point of view of environmental medicine and those that are considered by occupational medicine. This applies to substances such as phthalates, organophosphates, bisphenol A and TCDD, for which methods are presented in this edition.

In recent years the media have shown increasing interest in human biomonitoring. The fact that each of us is exposed to many substances is not new, but this has been spectacularly demonstrated by human biomonitoring in the case of some prominent personalities. Now various national and international committees have expressed the fear that the analytical capabilities of human biomonitoring may have outstripped our ability to interpret the resulting data. This cannot be otherwise, as data must first be gathered before they can be interpreted. It is important only that these data are accurate, and the methods published here are intended to further this aim.

In the light of this discussion I would like to thank the Deutsche Forschungsgemeinschaft, which began to sponsor this collection of methods more than 30 years ago. Although in those days it was no more than a justified hope that measurement of pollutants in the body fluids of humans would prove more reliable than other pro-

Foreword

cedures for detecting the absorbed doses and for reducing them if possible. Thanks to the support of the DFG, human biomonitoring has become an effective tool for the prevention of damage to health that requires no further legitimisation. I extend my personal thanks to Dr. Armin Krawisch from the business office of the DFG for his constant support, special thanks are also due to the members and guests of the "Analyses of Hazardous Substances in Working Materials" working group who are the mainstay of this publication. Moreover, I wish to thank Mr. Oliver Midasch and Ms. Barbara Müller-Göthert who also lend their reliable support to my work.

> Prof. Dr. Jürgen Angerer
> Chairman of the "Analytical Chemistry" working group of the Deutsche Forschungsgemeinschaft's Commission for the Investigation of Health Hazards of Chemical Compounds in the Work Area

Contents

Contents of Volumes 1–11 XI

Working Group Analytical Chemistry of the Commission of the Deutsche Forschungsgemeinschaft for the Investigation of Health Hazards of Chemical Compounds in the Work Area XXXIV

Organization – Objectives and operational procedure – Development, examination, release and quality of the analytical methods – Publications of the working group – Withdrawal of methods

Terms and symbols used XXXVI

Terminology – symbols

Preliminary Remarks

The use of liquid chromatography/mass spectrometry (LC/MS) in biological monitoring 3

Methods

Bisphenol A in urine 55
Di(2-ethylhexyl) phthalate (DEHP) metabolites (2-ethyl-5-hydroxyhexyl phthalate, 2-ethyl-5-oxohexyl phthalate, mono(2-ethylhexyl) phthalate) in urine 73
5-Hydroxy-N-methyl-2-pyrrolidone (5-HNMP) and **2-hydroxy-N-methylsuccinimide (2-HMSI)** in urine as metabolites of N-methyl-2-pyrrolidone (NMP) 97
Iridium in urine; addendum to "platinum and gold" 115
Monohydroxybutenylmercapturic acid (MHBMA) and **dihydroxybutylmercapturic acid (DHBMA)** in urine as metabolites of 1,3-butadiene 127
Neuropathy target esterase (NTE) in leukocytes 151
Organophosphates (chlorpyrifos, diazinon, fenitrothion, fenthion, malathion) in whole blood 167
Palladium in urine 189
Polycyclic musk compounds (PMC) in blood (1,3,4,6,7,8-hexahydro-4,6,6,7,8,8-hexamethy lcyclopenta[g]-2-benzopyrane (HHCB), 7-acetyl-1,1,3,4,4,6-hexamethyltetrahydro naphthalene (AHTN), 4-acetyl-1,1-dimethyl-6-tert-butylindane (ADBI) and 6-acetyl-1,1,2,3,3,5-hexamethylindane (AHDI)) 209

The MAK-Collection Part IV: Biomonitoring Methods, Vol. 11.
DFG, Deutsche Forschungsgemeinschaft
Copyright © 2008 WILEY-VCH Verlag GmbH & Co. KGaA, Weinheim
ISBN: 978-3-527-31596-3

Propylene and diethylene glycol ethers in urine and blood (1-butoxy-propanol-2, diethylene glycol dibutyl ether, diethylene glycol diethyl ether, diethylene glycol dimethyl ether, diethylene glycol monobutyl ether, diethylene glycol monobutyl ether actetate, diethylene glycol monoethyl ether, diethylene glycol monoethyl ether acetate, diethylene glycol monomethyl ether, dipropylene glycol monomethyl ether, 1-ethoxypropanol-2, 3-ethoxypropanol-1, 1-ethoxypropyl acetate-2, 1-methoxypropanol-2, 1-methoxypropanone-2, 1-methoxypropyl acetate-2, propylene glycol diacetate) 231

2,3,7,8-Tetrachlorodibenzo-p-dioxin in blood 261

Members and Guests of the Working Subgroup Analyses of Hazardous Substances in Biological Materials of the Commission of the Deutsche Forschungsgemeinschaft for the Investigation of Health Hazards of Chemical Compounds in the Work Area 279

Contents of Volumes 1–11

Substance	Vol.	Page
Acetone	see alcohols and ketones	
Acetylcholinesterase (AchE; acetylcholine-acetylhydrolase EC 3.1.1.7) in erythrocytes and cholinesterase (ChE: acylcholin-acylhydrolase EC 3.1.1.8) in plasma	3	45
Acetylcholine-acetylhydrolase	see acetylcholinesterase and cholinesterase	
N-Acetyltransferase 2 (genotyping)	9	135
N-Acetyltransferase 2 (phenotyping)	9	165
AChE	see acetylcholinesterase and cholinesterase	
Acrylnitrile	see N-2-cyanoethylvaline, N-2-hydroxyethylvaline, N-methylvaline	
Acylcholin-acylhydrolase	see acetylcholinesterase and cholinesterase	
ADBI	see polycyclic musk compounds	
AHDI	see polycyclic musk compounds	
AHTN	see polycyclic musk compounds	
Alcohols and ketones (acetone; 1-butanol; 2-butanol; 2-butanone; ethanol; 2-hexanone; methanol; 2-methyl-1-propanol; 4-methyl-2-pentanone; 1-propanol, 2-propanol) in blood and urine	5	1
Alkoxycarboxylic acids in urine as metabolites of glycol ethers with a primary alcohol group	10	55
Aluminium, chromium, cobalt, copper, manganese, molybdenum, nickel, vanadium in urine	7	73
Aluminium in plasma	6	47

Substance	Vol.	Page
Aminodinitrotoluenes in urine as metabolites of trinitrotoluene	10	81
2-Amino-4,6-dinitrotoluene	see aminodinitrotoluenes	
4-Amino-2,6-dinitrotoluene	see aminodinitrotoluenes	
4-Aminodiphenyl	see aromatic amines	
4-Aminodiphenyl	see haemoglobin adducts of aromatic amines	
Aminotoluenes	see haemoglobin adducts of aromatic amines	
Amitrole (3-amino-1,2,4-triazole) in urine . . .	6	63
Aniline	see aromatic amines	
Aniline	see haemoglobin adducts of aromatic amines	
o-anisidine	see aromatic amines	
Anthracycline cytostatic agents (doxorubicin, epirubicin, daunorubicin, idarubicin) in urine .	7	119
Antimony in blood and urine	2	31
Antimony in urine	4	51
Antimony	see ICP-MS collective method	
Application of the ICP-MS for biological monitoring	6	1
Aromatic alcohols	see phenols and aromatic alcohols	
Aromatic amines in urine (1-naphthylamine; 2-naphthylamine; 4,4'-methylene-bis(2-chloroaniline); 3,3'-dichlorobenzidine)	1	17
Aromatic amines (aniline; *o*-toluidine; *m*-toluidine; *p*-toluidine; 2,4- and 2,6-toluylenediamine; 4-aminodiphenyl; 4,4'-diaminodiphenylmethane) in urine, plasma and erythrocytes	4	67

Substance	Vol.	Page
Aromatic carboxylic acids in urine (phenylglyoxylic acid; mandelic acid; hippuric acid; o-methylhippuric acid, m-/p-methylhippuric acids; benzoic acid)	2	47
Arsenic in urine	3	63
Arsenic Species (As(III), As(V), monomethylarsonic acid, dimethylarsinic acid in urine	7	97
As(III)	see arsenic species	
As(V)	see arsenic species	
Barium in urine	3	81
Barium, strontium, titanium in urine	2	67
Benzene	see benzene and alkylbenzenes	
Benzene	see furan-2-carboxylic acid and other carboxylic acids	
Benzene	see t,t-muconic acid	
Benzene	see S-phenylmercapturic acid	
Benzene and alkylbenzenes (BTX-aromatics) in blood	4	107
Benzene derivatives in urine, suitable for steam distillation (phenol; m-/p-cresol; o-cresol; o-chlorophenol; o-nitrophenol; p-chlorophenol; nitrobenzene; 1,2-dinitrobenzene; 2-chloro-5-methylphenol; 2,5-dichlorophenol; 3,4-dichlorophenol; 2,3-dinitrotoluene)	1	31
Benzidine	see haemoglobin adducts of aromatic amines	
Benzoic acid	see aromatic carboxylic acids	
Benzoic acid	see furan-2-carboxylic acid and other carboxylic acids	
Benzyl alcohol	see furan-2-carboxylic acid and other carboxylic acids	

Substance	Vol.	Page
Benzylchloride	see N-benzylvaline	
N-Benzylvaline after exposure to benzylchloride in blood	8	35
Beryllium in urine	1	57
Beryllium, lithium, vanadium, tungsten in urine	5	51
Bromide in plasma and serum	10	97
Beryllium in urine, standard addition procedure	5	35
Bismuth	see ICP-MS collective method	
Bisphenol A in urine	11	55
Bromide in urine	1	67
2-Bromo-2-chloro-1,1,1-trifluoroethane	see halogenated hydrocarbons	
1,3-Butadiene	see monohydroxybutenylmercapturic acid (MHBMA) and dihydroxybutylmercapturic acid (DHBMA)	
1-Butanol	see alcohols and ketones	
2-Butanol	see alcohols and ketones	
2-Butanone	see alcohols and ketones	
Butoxyacetic acid in urine	4	131
2-Butoxyacetic acid	see alkoxycarboxylic acids	
2-Butoxyethanol	see alkoxycarboxylic acids	
2-Butoxyethyl acetate	see alkoxycarboxylic acids	
1-Butoxypropanol-2	see propylene and diethylene glycol ethers	
Butyldiglycol	see 2-(2-butoxyethoxy)ethanol	
2-Butoxyethoxyacetic acid	see alkoxycarboxylic acids	

Substance	Vol.	Page
2-(2-butoxyethoxy)ethanol	see alkoxycarboxylic acids	
Cadmium in blood	1	79
Cadmium in urine	2	85
Cadmium	see ICP-MS collective method	
Carbon disulphide	see furan-2-carboxylic acid and other carboxylic acids	
Carboxyhemoglobin in blood	1	93
ChE	see acetylcholinesterase and cholinesterase	
Chlorinated aromatic hydrocarbons in plasma (o-dichlorobenzene; m-dichlorobenzene; p-dichlorobenzene; 1,2,4-trichlorobenzene; 1,2,4,5-tetrachlorobenzene; pentachlorobenzene; hexachlorobenzene)	3	93
4-chloroaniline	see haemoglobin adducts of aromatic amines	
p-chloroaniline	see haemoglobin adducts of aromatic amines	
Chlorobenzenes in blood (1,2- and 1,4-dichlorobenzene; 1,2,4-trichlorobenzene)	1	107
2-Chloro-5-methylphenol	see benzene derivatives	
4-Chloro-2-methylphenoxyacetic acid	see chlorophenoxycarboxylic acids	
4-Chloro-2-methylphenoxypropionic acid	see chlorophenoxycarboxylic acids	
o-Chlorophenol, p-chlorophenol	see benzene derivatives	
Chlorophenols in urine (2,4-dichlorophenol, 2,5-dichlorophenol, 2,6-dichlorophenol, 2,3,4-trichlorophenol, 2,4,5-trichlorophenol, 2,4,6-trichlorophenol, 2,3,4,6-tetrachlorophenol)	7	143

Substance	Vol.	Page
4-Chloro-*o*-toluidine	see aromatic amines	
Chlorophenols (monohydroxychlorobenzenes) in urine (2,6-, 2,3-, 3,4-dichlorophenol; 2,4,6-, 2,4,5-, 3,4,5-trichlorophenol; 2,3,4,6-, 2,3,4,5-tetrachlorophenol; pentachlorophenol)	1	123
Chlorophenoxycarboxylic acids (4-chloro-2-methylphenoxyacetic acid; 2,4-dichlorophenoxyacetic acid; 4-chloro-2-methylphenoxypropionic acid; 2,4-dichlorophenoxypropionic acid) in urine	5	77
Chlorpyrifos	see organophosphates	
Chlorpyrifos	see 3,5,6-trichloro-2-pyridinol in urine	
Chlorpyrifos-methyl	see 3,5,6-trichloro-2-pyridinol in urine	
Cholinesterase	see acetylcholinesterase and cholinesterase	
Chromium	see aluminium, chromium, cobalt, copper, manganese, molybdenum, nickel, vanadium	
Chromium in urine	2	97
Chromium in whole blood, plasma and erythrocytes	3	109
Cobalt	see aluminium, chromium, cobalt, copper, manganese, molybdenum, nickel, vanadium	
Cobalt in urine	1	141
Cobalt in blood	2	117

Substance	Vol.	Page
Copper	see aluminium, chromium, cobalt, copper, manganese, molybdenum, nickel, vanadium	
Cotinine in urine, plasma or serum	7	171
Cotinine in urine	8	53
o-Cresol; *m-/p*-cresol	see benzene derivatives	
Cyanide in blood	2	133
N-2-Cyanoethylvaline, N-2-hydroxyethylvaline, N-methylvaline in blood	5	211
Cyclophosphamide	see oxazaphosphorines	
Cytochrome P450 1A1 (genotyping)	9	67
Cytochrome P450 1A1	see short guidelines for real time PCR	
Cytochrome P450 1B1 (genotyping)	9	89
Cytochrome P450 2E1 (genotyping)	9	111
Cytochrome P450 2E1	see short guidelines for real time PCR	
Daunorubicin	see anthracycline cytostatic agents	
DDE	see organochlorine compounds	
DDT	see organochlorine compounds	
DEHP	see di(2-ethylhexyl) phthalate (DEHP) metabolites	
DHBMA	see monohydroxybutenylmercapturic acid (MHBMA) and dihydroxybutylmercapturic acid (DHBMA)	
2,4-Diamino-6-chloro-s-triazine in urine	6	111

Substance	Vol.	Page
4,4′-Diaminodiphenylmethane	see aromatic amines	
4,4′-Diaminodiphenylmethane	see haemoglobin adducts of aromatic amines	
Diazinon	see organophosphates	
Dibenzodioxins	see dioxins, furans and WHO PCB	
Dibenzofurans	see dioxins, furans and WHO PCB	
cis-3-(2,2-dibromovinyl)-2,2-dimethylcyclopropane-1-carboxylic acid	see pyrethroid metabolites	
1,2-Dichlorobenzene	see 3,4-Dichlorocatechol and 4,5-dichlorocatechol	
1,2-; 1,4-Dichlorobenzene	see chlorobenzenes	
o-dichlorobenzene, m-dichlorobenzene, p-dichlorobenzene	see chlorinated aromatic hydrocarbons	
3,3′-Dichlorobenzidine	see aromatic amines	
3,3′-Dichlorobenzidine	see haemoglobin adducts of aromatic amines	
3,5-Dichloroaniline	see vinclozolin	
3,4-Dichlorocatechol and **4,5-dichlorocatechol** in urine	10	113
4,4′-Dichlorodiphenyldichloroethane	see organochlorine compounds	
4,4′-Dichlorodiphenyltrichloroethane	see organochlorine compounds	
1,2-Dichloroethylene	see chlorinated aromatic hydrocarbons	
Dichloromethane	see chlorinated aromatic hydrocarbons	
2,5-; 3,4-Dichlorophenol	see benzene derivatives	
2,6-; 2,3-; 3,4-Dichlorophenol	see chlorophenols	

Substance	Vol.	Page
2,4-Dichlorophenol	see chlorophenols	
2,5-Dichlorophenol	see chlorophenols	
2,6-Dichlorophenol	see chlorophenols	
2,4-Dichlorophenoxyacetic acid	see chlorophenoxycarboxylic acids	
2,4-Dichlorophenoxypropionic acid	see chlorophenoxycarboxylic acids	
2,4-Dichlorotoluene	see furan-2-carboxylic acid and other carboxylic acids	
cis-3-(2,2-dichlorovinyl)-2,2-dimethylcyclopropane-1-carboxylic acid	see pyrethroid metabolites	
trans-3-(2,2-dichlorovinyl)-2,2-dimethylcyclopropane-1-carboxylic acid	see pyrethroid metabolites	
Diethylene glycol dibutyl ether, diethylene glycol diethyl ether, diethylene glycol dimethyl ether, diethylene glycol monobutyl ether, diethylene glycol monobutyl ether actetate, diethylene glycol monoethyl ether, diethylene glycol monoethyl ether acetate, diethylene glycol monomethyl ether	see propylene and diethylene glycol ethers	
Di(2-ethylhexyl) phthalate (DEHP) metabolites (2-ethyl-5-hydroxyhexyl phthalate, 2-ethyl-5-oxohexyl phthalate, mono(2-ethylhexyl) phthalate) in urine	11	73
Digestion procedures for the determination of metals in biological material	2	1
Digestion procedures for the determination of metals in biological materials	8	1
Dihydroxybutylmercapturic acid	see monohydroxybutenylmercapturic acid (MHBMA) and dihydroxybutylmercapturic acid (DHBMA)	
3,4-Dihydroxychlorobenzene (4-chlorocatechol) in urine	6	125

Substance	Vol.	Page
N,N-Dimethylacetamide (DMA) and N-methylacetamide (NMA) in urine	8	67
Dimethylarsinic acid	see arsenic species	
N,N-Dimethylformamide (DMF) in urine ...	5	97
2,4-Dimethylnitrobenzene	see furan-2-carboxylic acid and other carboxylic acids	
2,4-; 2,3-; 3,4-Dimethylphenol	see phenols and aromatic alcohols	
1,2-Dinitrobenzene	see benzene derivatives	
o-dinitrobenzene	see nitroaromatic compounds	
2,3-Dinitrotoluene	see benzene derivatives	
2,6-Dinitrotoluene	see nitroaromatic compounds	
Dioxins, furans and WHO PCB in whole blood	8	85
Dipropylene glycol monomethyl ether	see propylene and diethylene glycol ethers	
Doxorubicin	see anthracycline cytostatic agents	
Epirubicin	see anthracycline cytostatic agents	
Erythrocyte porphyrins (free) in blood (erythrocytes)	2	145
Ethanol	see alcohols and ketones	
2-Ethoxyacetic acid	see alkoxycarboxylic acids	
2-Ethoxyethanol	see alkoxycarboxylic acids	
2-Ethoxyethyl acetate	see alkoxycarboxylic acids	
1-Ethoxypropanol-2, 3-ethoxypropanol-1, 1-ethoxypropyl acetate-2	see propylene and diethylene glycol ethers	
Ethylbenzene	see benzene and alkyl benzenes	

Substance	Vol.	Page
Ethylbenzene	see furan-2-carboxylic acid and other carboxylic acids	
Ethyl bromide	see bromide	
Ethylene oxide	see N-2-cyanoethylvaline, N-2-hydroxyethylvaline, N-methyl-valine	
2-Ethyl-5-hydroxyhexyl phthalate	see di(2-ethylhexyl) phthalate (DEHP) metabolites	
2-Ethyl-5-oxohexyl phthalate	see di(2-ethylhexyl) phthalate (DEHP) metabolites	
2-Ethylphenol	see phenols and aromatic alcohols	
Evaluation of susceptibility parameters in occupation and environmental medicine	9	315
Fenitrothion	see organophosphates	
Fenthion	see organophosphates	
Fluoride in urine	2	159
4-Fluoro-3-phenoxybenzoic acid	see pyrethroid metabolites	
Free erythrocyte porphyrins	see erythrocyte porphyrins (free)	
Furan-2-carboxylic acid and other carboxylic acids (phenylglyoxylic acid, mandelic acid, t,t-muconic acid, benzoic acid, hydroxybenzoic acid, hippuric acid, methylhippuric acid, 2,4-dichlorobenzoic acid, 3-methyl-4-nitro-benzoic acid, TTCA)	10	129
Furans	see dioxins, furans and WHO PCB	
Furfural	see 2-furylmethanal	
2-Furylmethanal	see furan-2-carboxylic acid and other carboxylic acids	
Gas chromatographic methods for the determination of organic substances in biological material	3	1

Substance	Vol.	Page
Glucose-6-phosphate dehydrogenase (G-6-PDH) (genotyping)	9	265
Glutathion S-transferase M1	see glutathion S-transferase T1 and M1 see short guidelines for real time PCR	
Glutathion S-transferase P1 (GSTP1) (genotyping)	9	221
Glutathion S-transferase P1	see short guidelines for real time PCR	
Glutathion S-transferase T1 (phenotyping) ..	9	211
Glutathion S-transferase T1	see short guidelines for real time PCR	
Glutathion S-transferase T1 and M1 (**GSTT1, GSTM1**) (genotyping)	9	183
Glycol ethers, glycol ether acetates	see propylene and diethylene glycol ethers	
Gold	see platinum and gold	
Haemoglobin adducts of aromatic amines (aniline, o-, m- and p-toluidine, o-anisidine, p-chloroaniline, α- and β-naphthylamine, 4-aminodiphenyl, benzidine, 4,4'-diaminodiphenylmethane, 3,3'-dichlorobenzidine)	7	191
Halogenated hydrocarbons in blood (dichloromethane; 1,2-dichloroethylene; 2-bromo-2-chloro-1,1,1-trifluoroethane (halothane); trichloromethane; 1,1,1-trichloroethane; tetrachloromethane; trichloroethylene; tetrachloroethylene)	3	127
Halothane	see halogenated hydrocarbons	
Hexachlorobenzene	see chlorinated aromatic hydrocarbons and oxachlorine compounds	
Hexachlorocyclohexane	see organochlorine compounds	

Substance	Vol.	Page
Hexamethylenediamine	see hexamethylene diisocyanate (HDI) and hexamethylenediamine	
Hexamethylene diisocyanate (HDI) and hexamethylenediamine (HDA) in urine	8	119
2,5-Hexanedione	see hexane metabolites	
Hexane metabolites (2,5-hexanedione, 2-hexanone) in urine	4	147
2-Hexanone	see hexane metabolites	
2-Hexanone	see alcohols and ketones	
HHCB	see polycyclic musk compounds	
Hippuric acid	see aromatic carboxylic acids and o-, m-/p-methylhippuric acids	
Hippuric acid	see furan-2-carboxylic acid and other carboxylic acids	
2-HMSI	see 5-hydroxy-N-methyl-2-pyrrolidone (5-HNMP) and 2-hydroxy-N-methylsuccinimide (2-HMSI)	
5-HNMP	see 5-hydroxy-N-methyl-2-pyrrolidone (5-HNMP) and 2-hydroxy-N-methylsuccinimide (2-HMSI)	
Hydrazine in blood (plasma)	2	171
Hydrazine and **N-acetylhydrazine** in urine and plasma	6	141
Hydroxybenzoic acid	see furan-2-carboxylic acid and other carboxylic acids	
8-Hydroxy-2′-deoxyguanosine in urine	8	133
1-(4-(1-Hydroxy-1-methylethyl)-phenyl)-3-methylurea (HMEPMU)	8	151

Substance	Vol.	Page
5-Hydroxy-*N*-methyl-2-pyrrolidone (5-HNMP) and **2-hydroxy-*N*-methylsuccinimide (2-HMSI)** in urine as metabolites of *N*-methyl-2-pyrrolidone (NMP)	11	97
1-Hydroxyphenanthrene	see PAH metabolites	
4-Hydroxyphenanthrene	see PAH metabolites	
9-Hydroxyphenanthrene	see PAH metabolites	
1-Hydroxypyrene in urine	3	151
1-Hydroxypyrene in urine	see PAH metabolites	
ICP-MS collective method (antimony, bismuth, cadmium, lead, mercury, platinum, tellurium, thallium, tin, tungsten) in urine	6	79
Idarubicin	see anthracycline cytostatic agents	
Ifosfamide	see oxazaphosphorines	
Indium in urine	3	171
Iridium in urine; addendum to "platinum and gold"	11	115
Isoproturon	see 1-(4-(1-hydroxy-1-methylethyl)-phenyl)-3-methylurea (HMEPMU)	
Lead in blood	1	155
Lead in blood and urine	2	183
Lead	see ICP-MS collective method	
Lithium	see beryllium, lithium, vanadium, tungsten	
Malathion	see organophosphates	
Mandelic acid	see aromatic carboxylic acids	
Mandelic acid	see furan-2-carboxylic acid and other carboxylic acids	

Substance	Vol.	Page
Manganese	see aluminium, chromium, cobalt, copper, manganese, molybdenum, nickel, vanadium	
Manganese in blood	10	157
MEHP	see di(2-ethylhexyl) phthalate (DEHP) metabolites	
Mercury in blood and urine	2	195
Mercury	see ICP-MS collective method	
Metasystox R in urine	7	221
Methanol	see alcohols and ketones	
Methoxyacetic acid	see alkoxycarboxylic acids	
Methoxyethanol	see alkoxycarboxylic acids	
Methoxyethyl acetate	see alkoxycarboxylic acids	
1-Methoxypropanol-2	see propylene and diethylene glycol ethers	
2-Methoxypropanol	see alkoxycarboxylic acids	
1-Methoxypropanone-2	see propylene and diethylene glycol ethers	
2-Methoxypropionic acid	see alkoxycarboxylic acids	
1-Methoxypropyl acetate-2	see propylene and diethylene glycol ethers	
2-Methoxypropyl acetate-1	see alkoxycarboxylic acids	
N-Methylacetamide	see N,N-dimethylacetamide (DMA) and N-methylacetamide (NMA)	
Methylating agents (bis(chloromethyl)ether, bromomethane, chloromethane, dimethyl sulphate, iodomethane, monochlorodimethyl ether)	see N-2-cyanoethylvaline, N-2-hydroxyethylvaline, N-2-methylvaline	

Substance	Vol.	Page
3-Methylbenzyl alcohol	see phenols and aromatic alcohols	
Methyl bromide	see bromide	
4,4′-Methylene-*bis*(2-chloroaniline)	see aromatic amines	
Methylhippuric acid	see furan-2-carboxylic acid and other carboxylic acids	
o-Methylhippuric acid, *m-/p*-methylhippuric acids (Toluric acids), hippuric acid in urine ..	1	165
o-Methylhippuric acid, *m-/p*-methylhippuric acids	see aromatic carboxylic acids	
Methylmercury in blood	10	169
3-Methyl-4-nitro-benzoic acid	see furan-2-carboxylic acid and other carboxylic acids	
4-Methyl-2-pentanone	see alcohols and ketones	
2-, 4-Methylphenol	see phenols and aromatic alcohols	
2-Methyl-1-propanol	see alcohols and ketones	
N-Methyl-2-pyrrolidone (NMP)	see 5-hydroxy-*N*-methyl-2-pyrrolidone (5-HNMP) and 2-hydroxy-*N*-methylsuccinimide (2-HMSI)	
MHBMA	see monohydroxybutenylmercapturic acid (MHBMA) and dihydroxybutylmercapturic acid (DHBMA)	
β_2-Microglobulin in urine and serum	3	185
Molybdenum	see aluminium, chromium, cobalt, copper, manganese, molybdenum, nickel, vanadium	
Molybdenum in urine	5	109
Molybdenum in plasma and urine	8	167
Mono(2-ethylhexyl) phthalate	see di(2-ethylhexyl) phthalate (DEHP) metabolites	

Substance	Vol.	Page
Monohydroxybutenylmercapturic acid (MHBMA) and **dihydroxybutylmercapturic acid (DHBMA)** in urine as metabolites of 1,3-butadiene	11	127
Monohydroxychlorobenzenes	see chlorophenols	
Monomethylarsonic acid	see arsenic species	
t,t-Muconic acid	see furan-2-carboxylic acid and other carboxylic acids	
***t,t*-Muconic acid** in urine	5	125
Musk compounds	see polycyclic musk compounds	
1-, 2-Naphthylamine	see aromatic amines	
1-Naphthylamine	see haemoglobin adducts of aromatic amines	
2-Naphthylamine	see haemoglobin adducts of aromatic amines	
α-Naphthylamine	see haemoglobin adducts of aromatic amines	
β-Naphthylaminex	see haemoglobin adducts of aromatic amines	
Neuropathy target esterase (NTE) in leukocytes	11	151
Nickel	see aluminium, chromium, cobalt, copper, manganese, molybdenum, nickel, vanadium	
Nickel in blood	3	193
Nickel in urine	1	177
Nitroaromatic compounds in plasma (nitrobenzene; *p*-nitrotoluene, *p*-nitrochlorobenzene; 2,6-dinitrotoluene; *o*-dinitrobenzene; 1-nitronaphthalene; 2-nitronaphthalene; 4-nitrobiphenyl)	3	207

Substance	Vol.	Page
Nitrobenzene	see benzene derivatives	
Nitrobenzene	see nitroaromatic compounds	
4-Nitrobiphenyl	see nitroaromatic compounds	
p-Nitrochlorobenzene	see nitroaromatic compounds	
1-, 2-Nitronaphthaline	see nitroaromatic compounds	
o-Nitrophenol	see benzene derivatives	
p-Nitrotoluene	see nitroaromatic compounds	
NMP	see 5-hydroxy-N-methyl-2-pyrrolidone (5-HNMP) and 2-hydroxy-N-methylsuccinimide (2-HMSI)	
NTE	see neuropathy target esterase (NTE)	
Organochlorine compounds in whole blood and plasma	8	187
Organophosphates in whole blood (chlorpyrifos, diazinon, fenitrothion, fenthion, malathion)	11	167
Organotin compounds (except methyltin compounds) and total tin in urine	4	165
Oxazaphosphorines: Cyclophosphamide and ifosfamide in urine	8	221
Oxydemeton-methyl in urine	see metasystox R	
PAH metabolites in urine	6	163
Palladium in urine	11	189
Pentachlorobenzene	see chlorinated aromatic hydrocarbons	
Pentachlorophenol	see chlorophenols	
Pentachlorophenol in urine	7	237
Pentachlorophenol in urine and serum/plasma	6	189

Substance	Vol.	Page
Perfluorooctanoic acid in plasma	10	191
Perfluorooctanesulphonic acid and Perfluorobutanesulphonic acid in plasma and urine	10	213
PFBS	see perfluorooctanesulphonic acid and perfluorobutane-sulphonic acid in plasma and urine	
PFOA	see perfluorooctanoic acid in plasma	
PFOS	see perfluorooctanesulphonic acid and perfluorobutane-sulphonic acid in plasma and urine	
Phenol in urine	1	189
Phenol	see benzene derivatives and phenols and aromatic alcohols	
Phenols and aromatic alcohols in urine (phenol; 2- and 4-methylphenol; DL-1- and 2-phenylethanol; 3-methylbenzyl alcohol; 2-ethylphenol; 2,4-; 2,3- and 3,4-dimethylphenol	2	213
Phenols in urine	6	211
Phenoxyacetic acid	see alkoxycarboxylic acids	
3-Phenoxybenzoic acid	see pyrethroid metabolites	
2-Phenoxyethanol	see alkoxycarboxylic acids	
DL-1-, 2-phenylethanol	see phenols and aromatic alcohols	
Phenylglyoxylic acid	see aromatic carboxylic acids	
Phenylglyoxylic acid	see furan-2-carboxylic acid and other carboxylic acids	
S-Phenylmercapturic acid in urine	5	143
Platinum in urine, blood, plasma/serum	4	187

Substance	Vol.	Page
Platinum	see ICP-MS collective method	
Platinum and gold in urine	7	255
Polychlorinated biphenyls in blood or serum	3	231
Polychlorinated biphenyls	see dioxins, furans and WHO PCB and see organochlorine compounds	
Polychlorinated dibenzodioxines	see dioxins, furans and WHO PCB	
Polychlorinated dibenzofurans	see dioxins, furans and WHO PCB	
Polycyclic musk compounds (PMC) in blood (1,3,4,6,7,8-hexahydro-4,6,6,7,8,8-hexamethylcyclopenta[g]-2-benzopyrane (HHCB), 7-acetyl-1,1,3,4,4,6-hexamethyltetrahydro naphthalene (AHTN), 4-acetyl-1,1-dimethyl-6-tert-butylindane (ADBI) and 6-acetyl-1,1,2,3,3,5-hexamethylindane (AHDI))	11	209
Polymerase chain reaction and its application in occupational and environmental medicine	9	35
1-Propanol	see alcohols and ketones	
2-Propanol	see alcohols and ketones	
Propylene and diethylene glycol ethers in urine and blood (1-butoxypropanol-2, diethylene glycol dibutyl ether, diethylene glycol diethyl ether, diethylene glycol dimethyl ether, diethylene glycol monobutyl ether, diethylene glycol monobutyl ether actetate, diethylene glycol monoethyl ether, diethylene glycol monoethyl ether acetate, diethylene glycol monomethyl ether, dipropylene glycol monomethyl ether, 1-ethoxypropanol-2, 3-ethoxypropanol-1, 1-ethoxypropyl acetate-2, 1-methoxypropanol-2, 1-methoxypropanone-2, 1-methoxypropyl acetate-2, propylene glycol diacetate)	11	231
Propylene glycol diacetate	see propylene and diethylene glycol ethers	

Substance	Vol.	Page
Pyrethroid metabolites (*cis*-3-(2,2-dichlorovinyl)-2,2-dimethylcyclopropane-1-carboxylic acid, *trans*-3-(2,2-dichlorovinyl)-2,2-dimethylcyclopropane-1-carboxylic acid, *cis*-3-(2,2-dibromovinyl)-2,2-dimethylcyclopropane-1-carboxylic acid, 4-fluoro-3-phenoxybenzoic acid, 3-phenoxybenzoic acid) in urine	6	231
Rhodium in urine and serum/plasma	7	273
Selenium in blood, plasma and urine	2	231
Short guidelines for real-time PCR	9	281
Strontium	see barium, strontium, titanium	
Styrene	see furan-2-carboxylic acid and other carboxylic acids	
Sulphotransferase 1A1 and 1A2 (genotyping)	9	241
Susceptibility and biological monitoring	9	1
Susceptibility and determination methods	9	5
TCDD	see 2,3,7,8-tetrachlorodibenzo-p-dioxin	
TCPyr	see 3,5,6-trichloro-2-pyridinol in urine	
Tellurium	see ICP-MS collective method	
1,2,4,5-Tetrachlorobenzene	see chlorinated aromatic hydrocarbons	
2,3,7,8-Tetrachlorodibenzo-p-dioxin in blood	11	261
Tetrachloroethylene	see halogenated hydrocarbons	
Tetrachloromethane	see halogenated hydrocarbons	
2,3,4,6-; 2,3,4,5-Tetrachlorophenol	see chlorophenols	
Thallium in urine	1	199
Thallium in urine	5	163
Thallium	see ICP-MS collective method	

Substance	Vol.	Page
The use of atomic absorption spectrometry for the determination of metals in biological materials	4	1
The use of liquid chromatography/mass spectrometry (LC/MS) in biological monitoring	11	3
2-Thioxothiazolidine-4-carboxylic acid (TTCA) in urine	4	207
Thorium	see thorium and uranium	
Thorium and uranium in urine	6	255
Tin in urine	4	223
Tin	see ICP-MS collective method	
Titanium	see barium; strontium, titanium	
Titanium in blood and urine	10	237
Toluene	see benzene and alkylbenzenes	
Toluene	see furan-2-carboxylic acid and other carboxylic acids	
Toluidines	see haemoglobin adducts of aromatic amines	
m-Toluidine	see haemoglobin adducts of aromatic amines	
o-Toluidine	see haemoglobin adducts of aromatic amines	
p-Toluidine	see haemoglobin adducts of aromatic amines	
o-Toluidine, *m*-Toluidine, *p*-Toluidine	see aromatic amines	
Toluric acids	see methylhippuric acids	
2,4- and 2,6-Toluylendiamine	see aromatic amines	
Trichloroacetic acid (TCA) in urine	1	209

Substance	Vol.	Page
1,2,4-Trichlorobenzene	see chlorobenzenes and chlorinated aromatic hydrocarbons	
Trichloroethane	see halogenated hydrocarbons	
Trichloroethylene	see halogenated hydrocarbons	
Trichloromethane	see halogenated hydrocarbons	
2,4,6-; 2,4,5-; 3,4,5-Trichlorophenol	see chlorophenols	
2,3,4-Trichlorophenol	see chlorophenol	
2,4,5-Trichlorophenol	see chlorophenol	
2,4,6-Trichlorophenol	see chlorophenol	
3,5,6-Trichloro-2-pyridinol in urine	10	253
Trinitrotoluene	see aminodinitrotoluenes	
Tungsten	see beryllium, lithium, vanadium, tungsten and ICP-MS collective method	
Uranium	see thorium and uranium	
Vanadium	see aluminium, chromium, cobalt, copper, manganese, molybdenum, nickel, vanadium	
Vanadium in urine	3	241
Vanadium	see beryllium, lithium, vanadium, tungsten	
Vinclozolin as 3,5-dichloroaniline in urine . . .	7	287
Xylene	see furan-2-carboxylic acid and other carboxylic acids	
o-xylene, *m*-xylene	see benzene and alkylbenzenes	
Zinc in plasma, serum and urine	5	211

Working Group Analytical Chemistry of the Commission of the Deutsche Forschungsgemeinschaft for the Investigation of Health Hazards of Chemical Compounds in the Work Area

Organization

The Working Group Analytical Chemistry (Chairman: J. Angerer) was established in 1969. It includes two Working Subgroups: Air Analyses (Leader: H. Parlar) and Analyses of Hazardous Substances in Biological Materials (Leaders: J. Angerer, K. H. Schaller). The participants, who have been invited to collaborate on a Working Subgroup by the leaders, are experts in the field of technical and medical protection against chemical hazards at the workplace.

A list of the members and guests of Analyses of Hazardous Substances in Biological Materials is given at the end of this volume.

Objectives and operational procedure

The two analytical subgroups are charged with the task of preparing methods for the determination of hazardous industrial materials in the air of the work place or to determine these hazardous materials or their metabolic products in biological specimen from the persons working there. Within the framework of the existing laws and regulations, these analytical methods are useful for ambient monitoring at the workplace and biological monitoring of the exposed persons.

In addition to working out the analytical procedure, these subgroups are concerned with the problems of the preanalytical phase (specimen collection, storage, transport), the statistical quality control, as well as the interpretation of the results.

Development, examination, release and quality of the analytical methods

In its selection of suitable analytical methods, the Working Group is guided mainly by the relevant scientific literature and the expertise of the members and guests of the Working Subgroup. If appropriate analytical methods are not available they are worked out within the Working Group. The leader designates an author, who assumes the task of developing and formulating a method proposal. The proposal is examined experimentally by at least one other member of the project, who then submits a written report of the results of the examination. As a matter of principle the examination must encompass all phases of the proposed analytical procedure.

The examined method is then laid before the members of the subgroup for consideration. After hearing the judgement of the author and the examiner they can approve the method. The method can then be released for publication after a final meeting of the leader of the Working Group Analytical Chemistry with the subgroup leaders, authors, and examiners of the method.

Under special circumstances an examined method can be released for publication by the leader of the Working Group after consultation with the subgroup leaders.

Only methods for which criteria of analytical reliability can be explicitly assigned are released for publication. The values for inaccuracy, imprecision, detection limits, sensitivity, and specifity must fulfil the requirements of statistical quality control as well as the specific standards set by occupational health. The above procedure is meant to guarantee that only reliably functioning methods are published, which are not only reproducible within the framework of the given reliability criteria in different laboratories but also can be monitored over the course of time.

In the selection and development of a method for determining a particular substance the Working Group has given the analytical reliability of the method precedence over aspects of simplicity and economy.

Publications of the working group

Methods released by the Working Group are published in Germany by the Deutsche Forschungsgemeinschaft as a loose-leaf collection entitled "Analytische Methoden zur Prüfung gesundheitsschädlicher Arbeitsstoffe" (Wiley-VCH, Weinheim, Germany). The collection at present consists of two volumes:

Volume I: Luftanalysen
Volume II: Analysen in biologischem Material

These methods are published in an English edition. The work at hand represents the 11th English issue of "Biomonitoring Methods", formerly "Analyses of Hazardous Substances in Biological Materials".

Withdrawal of methods

An analytical method that is made obsolete by new developments or discoveries in the fields of instrumental analysis or occupational health and toxicology can be replaced by a more efficient method. After consultation with the membership of the relevant project and with the consent of the leader of the Working Group, the subgroup leader is empowered to withdraw the old method.

Terms and symbols used

Terminology

Accuracy – the agreement between the best estimate of a quantity and its true value. It has no numerical value.

Analyte – the component to be measured.

Assigned value – the concentration of the analyte in the control specimen assigned either arbitrarily, e.g., convention, or from preliminary evidence, e.g., in the absence of a recognized reference method.

Biological Tolerance Value for a Working Material (BAT) – is defined as the maximum permissible quantity of a chemical compound, its metabolites, or any deviation from the norm of biological parameters induced by these substances in exposed humans. According to current knowledge these conditions generally do not impair the health of the employee, even if exposure is repeated and of long duration. As with MAK values, the maximum period of exposure to a working material is generally given as eight hours daily and 40 hours weekly.

Blank value – the analytical result obtained when the complete procedure is carried out on ultrapure water containing no analyte instead of biological specimens.

Calibration standards – specimens with known concentrations of the analyte that are used for calibration.

Certified value – the concentration of the analyte in control specimens certified by an official body subject to conditions established by that body.

Control material – a material that is used solely for quality control purposes and not for calibration.

Definitive method – the analytical method of all the methods for determining the analyte that is capable of providing the highest accuracy. Its accuracy must be adequate for its stated purposes.

Definitive value – the concentration of analyte in control specimens determined by a definitive method. It is the best available estimate of the true value.

Detection limit – the minimum analytical result that is still clearly detectable and distinguishable from the background noise; defined as three standard deviations of the appropriate blank value.

External quality control – a procedure of utilizing for quality control purposes the results of analyses performed on the same specimen or specimens by several laboratories.

Imprecision – the standard deviation or coefficient of variation of the results in a set of replicate measurements. A distinction is made between within-series, between-day, and interlaboratory imprecision.

Inaccuracy – the numerical difference between the mean of a set of replicate measurements and the true value.

Influence factors – lead to changes in vivo in the clinical chemical parameter. Their influences are independent of the specificity of the analytical method.
Interference – the effect on the accuracy of measurement of one component caused by another component that does not itself produce a reading.
Interference factors – all factors that alter the result in vitro, i.e., after the specimen has been collected from the patient.
Internal quality control – the procedure of utilizing the results of only one laboratory for quality control purposes. It includes the control of imprecision as well as inaccuracy.

Maximum Concentration Value at the Workplace (MAK) – is defined as the maximum permissible concentration of a chemical compound present in the air within a working area (as gas, vapor particulate matter) which, according to current knowledge, generally does not impair the health of the employee nor cause undue annoyance. Under these conditions, exposure can be repeated and of long duration over a daily period of eight hours, constituting an average work week of 40 hours (42 hours per week as averaged over four successive weeks for firms having four work shifts).
Measure – the measured change of one parameter of the analyte in the physical or chemical system used for analysis.

Preanalytical phase – the period from the specimen collection to the aliquotation (sampling) of the biological specimens.
Precision – the agreement between replicate measurements. It has no numerical value.
Prognostic range – an interval that with a given probability includes the analytical result from an identical specimen.

Quality control – the study of those errors that are the responsibility of the laboratory and the procedures used to recognize and minimize them. This study includes all errors arising within the laboratory during the time from aliquotation of the specimen to dispatch of the report.

Recovery rate – in recovery experiments the amount of recovered analyte divided by the amount of added analyte expressed as a percentage.
Reference material – a material or substance for which one or more properties are sufficiently well established to be used for calibration of an apparatus or for verification of an analytical method.
Reference method – an analytical method whose inaccuracy and imprecision are small enough as demonstrated by direct comparison with the definitive method and show low incidence of susceptibility to known interferences is so well documented that the stated aims of the reference method may be achieved.
Reference method value – the most probable value derived from a set of results obtained by the most reliable reference methods available.
Reliability criteria – defined quantifiable parameters for the assessment of the quality of an analytical method, e.g., imprecision, inaccuracy, detection limit.

Terms and symbols used

Sample – that appropriate representation portion of a specimen used in the analysis.
Sensitivity – the differential quotient of the calibration function.
Selected method – routine method with known systematic error.
Specificity – the ability of an analytical method to determine solely the component or components it purports to measure. It has no numerical value.
Specimen – the material available for analysis.

Symbols

c	substance concentration of analyte
$\bar{c}$	mean substance concentration of analyte
E	*extinction*
k'	reciprocal calibration factor
m	mass
M	molar mass
n	number
p	pressure
P	probability
r	recovery rate
s	relative standard deviation
$s_{bl,abs}$	standard deviation of the blank value
s_d	relative standard deviation derived from duplicate analyses
s_w	relative standard deviation derived from replicate analyses of the same specimen
t_p	Student's t factor
T	thermodynamic temperature
u	prognostic range
V	volume
V_m	molar volume
x	observed measure of analyte
$\bar{x}$	mean
x_{bl}	observed measure of blank
z	number of duplicate analyses
H	sensitivity
ϱ	mass concentration
σ	volume concentration

Preliminary Remarks

The use of liquid chromatography/mass spectrometry (LC/MS) in biological monitoring

Abbreviations

ADS	Alkyl-diol-silica phase
AGP	α_1-acid glycoprotein
APCI	Atmospheric pressure chemical ionisation
API	Atmospheric pressure ionisation
APPI	Atmospheric pressure photo-ionisation
CE	Capillary electrophoresis
CEC	Capillary electrochromatography
CGE	Capillary gel electrophoresis
CIEF	Capillary isoelectric focusing
CNL	Constant neutral loss
CRM	Charge residue model
CZE	Capillary zone electrophoresis
DEHP	Di(2-ethylhexyl)phthalate
EPI	Enhanced product ion
ESI	Electrospray ionisation
FT-MS	Fourier transform mass spectrometer
GC/MS	Gas chromatography/mass spectrometry
HLB	Hydrophilic lipophilic balance
HPLC	High performance liquid chromatography
ICR-MS	Ion cyclotron resonance mass spectrometer
IEM	Ion evaporation model
ISRP	Internal surface reversed phase
ITP	Isotachophoresis
LC/MS	Liquid chromatography/mass spectrometry
LPS	Large particle support
MEKC	Micellar electrokinetic (capillary) chromatography
MRM	Multiple reaction monitoring
m/z	Mass-to-charge ratio
PAH	Polycyclic aromatic hydrocarbons
pCEC	Pseudo-capillary electrochromatography

The MAK-Collection Part IV: Biomonitoring Methods, Vol. 11.
DFG, Deutsche Forschungsgemeinschaft
Copyright © 2008 WILEY-VCH Verlag GmbH & Co. KGaA, Weinheim
ISBN: 978-3-527-31596-3

QqTOF	Hybrid quadrupole time-of-flight mass spectrometer
RAM	Restricted access media
RPC	Reverse phase chromatography
SIM	Selected ion monitoring
S/N	Signal-to-background noise ratio
SPE	Solid phase extraction
SPS	Semi-permeable surface
SRM	Selected reaction monitoring
TFC	Turbulent flow chromatography
TOF-MS	Time-of-flight mass spectrometer
UV	Ultraviolet

1 Introduction

Liquid chromatography/mass spectrometry (LC/MS) is a new and unique way of coupling the well-known analytical techniques of liquid chromatography and mass spectrometry. Starting with the pioneering research of Dole et al. [1] and Tal'roze et al. [2–4] at the end of the 1960s, this new technique achieved its greatest international scientific recognition to date with the award of the Nobel Prize for Chemistry to John B. Fenn in 2002.

Whereas gas chromatography/mass spectrometry (GC/MS) is one of the established routine techniques that is widely used for many different fields of application in analytical laboratories on account of the low purchasing price of the instruments, LC/MS is still at the stage of being introduced to the majority of potential users. LC/MS has sparked keen interest since it has proved successful in the determination of many organic compounds in the last twenty years, in particular polar, thermally labile substances and those with high molecular weights (>1000 u) that are difficult or impossible to analyse by means of GC/MS. It is estimated that approximately 80% of all the known organic compounds can be analysed using LC/MS, whereas only about 20% of all known organic species are accessible to GC/MS analysis without previous derivatisation [5]. In the last five years more than a dozen new LC/MS systems have become commercially available at reasonable prices. This has acted as an additional incentive to interested scientists to use LC/MS. It should be noted that basic research in this field has by no means been concluded. In addition to the steadily growing use of LC/MS systems for routine analysis, new instruments are still being developed and optimised for innovative applications.

When used for the purpose of biological monitoring (BM), LC/MS offers new opportunities in the fields of occupational and environmental medicine, especially for the specific and highly sensitive assay of polar metabolites of health-impairing working materials and xenobiotics as well as for the analysis of binding products (adducts) of hazardous substances with biological macromolecules (proteins and DNA).

As one of the most important techniques of liquid chromatography in analytical practice, high performance liquid chromatography (HPLC) has already been reviewed in a chapter in this collection of methods "Analyses of hazardous substances in biological materials" (General Introduction to Volume 7). "The use of gas chromatography-mass spectrometry in biological monitoring" (General Introduction to Volume 10 of this series, which was recently published) also provided a comprehensive description of the physico-chemical fundamental principles of mass spectrometry and reviewed the instruments that are currently in use.

This chapter on "The use of liquid chromatography/mass spectrometry (LC/MS) in biological monitoring" is intended to provide considerable supplementary information on the above-mentioned current advances to that contained in the chapter on "LC-MS-coupling" (General Introduction to Volume 7, Section 5.5). The practical and instrumental principles of LC/MS will be described as well as some aspects of method development and optimisation with examples of applications for special substance groups. Wherever appropriate, we will refer the reader to the previous publications (HPLC, Vol. 7) and (GC/MS, Vol. 10) to avoid repetition.

The "biomonitoring methods" series already contains one LC/MS method ("Perfluorooctanesulphonic acid and perfluorobutanesulphonic acid in plasma and urine", Volume 10); this present volume includes two further methods ("Di(2-ethylhexyl) phthalate (DEHP) metabolites in urine" and "Monohydroxybutenylmercapturic acid (MHBMA) and dihydroxybutylmercapturic acid (DHBMA) in urine").

2 Principle of liquid chromatography/mass spectrometry (LC/MS)

As in the case of GC/MS, two independently functioning analytical systems are coupled in LC/MS. As a rule, the analyte(s) of interest is/are separated from the sample matrix by means of the liquid chromatographic system (e.g. HPLC or capillary electrophoresis (CE)) and introduced into the mass spectrometer as ions. Compared with e.g. UV detection, mass spectrometric analysis offers such a wealth of specific information on a substance that it is generally possible to determine structurally different analytes even if they have identical retention times, and to quantify them separately if desired.

Whereas the analyte is already in the vaporous state after separation by gas chromatography in GC/MS and only has to be ionised, in LC/MS an additional phase change is required, i.e. from the analyte dissolved in a liquid to the analyte in the gaseous phase (see Figure 1).

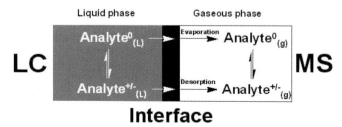

Fig. 1. Phase change and ionisation of an analyte in LC/MS according to [5]

These changes are achieved by specially designed interfaces. In general, LC/MS interfaces must be capable of facilitating three processes:

- evaporation of liquids and analyte molecules to vapours,
- ionisation of the uncharged analyte molecules in the gaseous phase or desorption of analyte ions from the liquid phase,
- evacuation of a strong stream of gas to maintain the high vacuum in the mass spectrometer.

Most LC/MS interfaces ensure that processes 1 and 2 proceed by the formation of an aerosol, as aerosol formation provides a highly effective enlargement of the surface area. The larger the surface area, the more readily the liquids and uncharged analyte molecules evaporate and the more readily analyte ions are desorbed from the liquid phase into the gaseous phase. Therefore improvement of the aerosol formation process poses a permanent challenge to manufacturers of LC/MS interfaces.

The third process, the evacuation of a strong stream of gas, poses a problem that can only be solved by great technical ingenuity, even in the case of GC/MS. Carrier gas flows of 0.5 to 2 mL/min are found there, which result in gas pressures of 1 to 2 bar at the end of the column. However, the use of a typical mass spectrometer requires a stable high vacuum of 10^{-5} to 10^{-6} mbar (see General Introduction to Volume 10 in this series). Overcoming the pressure difference when a liquid chromatographic system is coupled with a mass spectrometer presents an even greater challenge. As shown in Table 1, gas flows of more than 1000 mL/min may occur, depending on the type of LC column, the eluent(s) used and the flow rate. Only at very low flow rates (<10 µL/min) are the gas flow rates for the LC/MS coupling in the same order of magnitude as those for the carrier gas in a GC/MS coupling.

Technically the problem of the pressure difference at the LC/MS interfaces of practical importance is solved by installing several pumping stages in series to attain a high vacuum in the mass spectrometer. The pump technology used for creation of the vacuum has already been presented in the General Introduction to Volume 10 of this series.

Table 1. Comparison of (carrier) gas flow rates depending on the type of the GC or LC column, the carrier gas or eluent and its flow rate [6]

Column	Carrier gas/ Eluent	Flow rate [mL/min]	(Carrier) gas flow [mL/min]
GC (capillary)	Helium	1	1
GC (packed)/GC (CI)	Helium	20	20
LC (analytical)	Hexane	1	184
	Methanol	1	593
	Water	1	1240
LC (capillary)	Hexane	0.01	1.8
	Water	0.01	12

The use of liquid chromatography/mass spectrometry (LC/MS)

If all LC/MS systems are taken into account, they have the following set-up in common:

- a sample injection or chromatographic system
- an LC/MS interface
- one or more mass filters
- a detector

The individual components will be described in detail and their importance for LC/MS will be discussed in Section 3. Wherever appropriate, we have chosen to provide a brief description and will refer readers to the principles presented in the previously published chapters (HPLC, Vol. 7) and (GC/MS, Vol. 10) to avoid repetition.

3 Set-up of liquid chromatography/mass spectrometry coupling

3.1 Sample injection or chromatographic system

Three main techniques can be used for sample injection into an LC/MS system: a syringe pump, high performance liquid chromatography (HPLC) and capillary electrophoresis (CE). Independent of the technique used, sample injection must be performed as stipulated in the General Introduction to Vol. 7. It is particularly important to note that most LC/MS interfaces do not function well when samples contain high electrolyte concentrations (non-volatile salts, certain acids and alkalis) and matrix components. Under the most favourable conditions the performance of the instrument is impaired, and in the worst-case scenario severe damage to the relevant LC/MS interface may be caused. In this case it is advisable to precipitate proteins by the addition of organic solvents, as the sample is also simultaneously diluted. Alternatively, ultrafiltration may be carried out with subsequent dilution of the filtrate if necessary. Further currently employed procedures for sample processing are discussed in detail in Section 3.1.2.2 of this chapter.

3.1.1 Syringe pump

The syringe pump introduces the sample directly into the MS and, as a rule, it is supplied when the system is purchased. Flow rates from several nL/min up to several mL/min can be selected, depending on the size of syringe used, but 5 to 50 µL/min are normal. The syringe pump continuously infuses an analyte mixture onto the system during sample injection, but the individual components of a complex sample matrix are not separated off. In this case the specificity of the analysis is not based on the combination of a specific retention time with the results of the mass spectrometric analysis, but solely on the substance-specific information provided by the mass spectrometer. Therefore the syringe pump is suitable for two applications: the mass spectrometric analysis of reference substances (standards) and investigation of previously cleaned-up fractions.

In the first case the chemical structure is already known, generally by characterisation using another physico-chemical procedure (e.g. nuclear resonance spectroscopy). These substances serve as standards e.g. for the development of LC/MS methods that involve a chromatographic separation of the analyte(s) from the sample matrix. The pure analyte itself (insofar as its chemical structure is known) or its analogue labelled with heavy isotopes is often selected to serve as a standard ("internal standard"). Infusion of the solution containing the relevant pure substance with the aid of the syringe pump yields initial information on the behaviour of the reference substance (or that of the analyte) at the LC/MS interface and in the mass spectrometric analysis. The analyte ion yield can be qualitatively and quantitatively investigated and optimised by variation of the infusion conditions (e.g. flow rate, solvent composition of the sample) and by adjustment of the instrumental parameters for the LC/MS interface and mass spectrometer. The results obtained form the basis for the set-up of an "on-line" LC/MS procedure (see also Section 4.2 "Development and optimisation of methods" under "Tuning").

The second application is to facilitate the investigation of previously cleaned-up fractions. Such fractions can be obtained e.g. by liquid-liquid extraction, by solid phase extraction or in the form of an eluate from a preparative chromatographic procedure. In this case the specificity of the analysis is based on the selective enrichment of the analyte by extraction or chromatography and on the results of the mass spectrometric analysis. The "off-line" LC/MS methodology presented here has proved successful in clarifying the structure of unknown compounds as well as in measurement procedures with a high throughput in pharmaceutical analysis [7].

When a syringe pump is used, a technical distinction is made between low flow direct infusion and high flow infusion.

In the case of low flow direct infusion the syringe pump is directly connected to the LC/MS interface. This configuration is preferred for small sample volumes; selection of low flow rates from the nL/min range up to the lower µL/min range permits complete mass spectrometric analysis with high sensitivity, while sample consumption is kept to a minimum.

An LC system is required for infusion at higher flow rates. In this configuration the LC system is connected to the LC/MS interface by means of a T-junction. The syringe pump is connected to the second inlet on the T-junction. It is important to check that all the connections are tightly sealed, as even tiny leaks can have dramatic effects on the quality of the results.

The LC system is normally operated at a flow rate of 100 µL/min to 1 mL/min. The syringe pump infuses the sample solution into this constant stream of solvent (typical flow rates 1 to 10 µL/min). The resulting dilution of the sample solution depends on the ratio of the selected flow rates to each other. Good results can still be achieved in this configuration with concentrations of 1 pg/µL (after dilution).

Adverse effects on the LC/MS interface due to high electrolyte concentrations in the sample solution can be greatly reduced or even eliminated by dilution. However, some LC/MS interfaces achieve high ion yields only when flow rates are high, and an infusion at a high flow rate is then a prerequisite for correct analysis. Finally, the parameters of an on-line LC/MS method, such as different solvent compositions and different flow rates, can be simulated very well and varied rapidly in this configuration. The effects of each variation on the analyte ion yield in the LC/MS interface and the mass spectrometer can be investigated qualitatively and quantitatively. The information obtained from such experiments forms an important basis for setting up an on-line LC/MS method.

3.1.2 High performance liquid chromatography (HPLC)

High performance liquid chromatography (HPLC) is the most widely employed separation technique for routinely applied on-line LC/MS methods [5, 8]. The principles of HPLC have already been described in detail in this "Analyses of hazardous substances in biological materials" collection of methods (General Introduction to Volume 7). Therefore only special aspects of HPLC when combined with MS will be discussed here. We have chosen a general approach and in the following sections we will deal with some specific points, e.g. individual LC/MS interfaces in combination with HPLC.

3.1.2.1 MS-compatible HPLC systems

In theory all the HPLC phase systems mentioned in Volume 7, such as reverse phase chromatography, ion exchange chromatography and affinity chromatography, can be considered for use in an LC/MS configuration. In practice, however, their applicability is limited because compatibility with the relevant LC/MS interface is essential. Significant factors that influence the required compatibility are the flow rate at the end of the column and the composition of the eluents there, which, as stated above, may not contain high concentrations of electrolytes. In particular, reverse phase chromatography (RPC) meets these requirements and is therefore compatible with all known LC/MS interfaces. Therefore about 80% of all LC/MS methods with an HPLC application use RPC [5].

The inner diameter of the HPLC column has proved a useful parameter for classifying the HPLC columns used for LC/MS. Several groups can be distinguished: columns for conventional HPLC (preparative and analytical: wide-bore and normal-bore), narrow-bore columns and columns for capillary liquid chromatography (micro-bore and nano-bore). Table 2 shows typical inner diameters and flow rates.

Although preparative HPLC columns are included in this list for the sake of completeness, their importance for LC/MS is negligible. In contrast, analytical (normal-bore) and narrow-bore columns are prevalent in routinely used on-line LC/MS systems; there is also a growing tendency to use micro-bore and nano-bore columns for LC/MS applications [8]. Table 3 gives the most important characteristics of the various types of columns.

When the figures in Table 3 are considered, it is evident that miniaturisation of HPLC columns is accompanied by reduction in the amount of solvent consumed (smaller dead volume of the column, lower flow rates), less sample consumption

Table 2. Classification of HPLC columns for LC/MS according to their inner diameter [8]

Type of column	Inner diameter [mm]	Flow rate [mL/min]
Preparative (wide-bore)	>4.6	>3
Analytical (normal-bore)	3–4,6	0.5–3
Narrow-bore	1–2	0.02–0.3
Micro-bore	0.15–0.8	0.002–0.02
Nano-bore	0.02–0.1	0.0001–0.001

Table 3. Typical characteristics of HPLC columns according to Tomer et al. [9] and Abian et al. [8] (uniform column length: 25 cm)

Inner diameter [mm]	Column dead volume [µL]	Flow rate [mL/min]	Injection volume [µL]	Relative concentration at detector
4.6	4100	1	100	1
2	783	0.2	19	5.3
1	196	0.047	4.7	21.2
0.32	20	0.0049	0.485	206
0.05	0.49	0.00012	0.012	8459

(smaller injection volume) and a dramatic enhancement in the detection sensitivity (increase of the relative concentration at the detector by three powers of 10). These figures present clear arguments in favour of increasing the use of capillary liquid chromatography in LC/MS, which has not yet occurred in routine analysis to date.

However, technical and practical aspects must also be taken into consideration when HPLC columns are selected for LC/MS. For instance, analytical and narrow-bore columns can be operated with conventional HPLC systems, whereas existing systems have to be converted in order to perform capillary liquid chromatography [10, 11]. Change parts are commercially available to convert the common HPLC systems.

In addition, it should be noted that a method using an analytical or narrow-bore HPLC column that has already been established with another detector can generally be more rapidly modified to a LC/MS method. The already developed chromatography can often be taken over; the existing method must only be checked for compatibility with the LC/MS interface, and optimised if necessary.

Moreover, the quality of the HPLC columns used plays an important role. A large selection of analytical and narrow-bore columns packed with various materials is available from different commercial suppliers. They all provide a test chromatogram showing the most important chromatographic indices. This information indicates the batch conformity of the packing material and the compliance of its processing with requirements. The guarantee of conformity is of critical importance for an established LC/MS routine procedure that is set up for high sample throughput, as otherwise fluctuations in the quality of the packing material would entail time-consuming optimisation of the existing chromatography each time the column had to be exchanged.

In contrast to the analytical and narrow-bore HPLC columns, only a small selection of micro-bore and nano-bore columns are available from a few commercial suppliers [8]. One reason for the limited range is that the columns already available are undergoing continual development. Under these circumstances and due to the relatively high acquisition costs, many users have opted to prepare their own packed micro-bore and nano-bore columns according to various procedures [9, 12–17]. On the one hand, this has the advantage that the chromatography can be very rapidly and flexibly modified and optimised during development of an LC/MS method as part of research work. On the other, as discussed above, possible fluctuations in the quality of in-house packed HPLC columns may have negative effects on an LC/MS routine procedure once it has been established.

The technically possible injection volume also represents an important criterion for the choice of HPLC columns for LC/MS. Whereas analytical and narrow-bore HPLC columns may permit input of large sample volumes containing low analyte concentrations at µL levels using standard sample injection systems (General Introduction, Volume 7, Section 5.1), the injection volumes for capillary liquid chromatography are in the nL range. This requires the use of specially adapted sample input systems and prior enrichment of samples. Despite such difficulties various techniques [8] can be utilised to enrich the samples at the beginning of the column in order to enable µL-level sample volumes with low analyte concentrations to be injected onto micro-bore and nano-bore columns. However, overloading may easily result on account of the low maximum capacity of micro-bore and nano-bore columns, especially when the analyte(s) are contained in complex matrices. The use of a narrow-bore pre-column for on-line concentration of the sample and subsequent desorption of the enriched fraction for capillary liquid chromatographic analysis offers a possible solution [18, 19].

When the technical and practical aspects of the selection of HPLC columns for LC/MS are taken into consideration, it is understandable why many users [8] choose the analytical and narrow-bore HPLC column options: existing HPLC systems can be harnessed without changing the configuration, standard HPLC methods already established with other detectors can be readily adapted to LC/MS procedures, high quality, standardised column material is commercially available, and it is possible to inject large sample volumes with low concentrations of the analyte in complex matrices without problems using standard sample injection systems. If we compare the frequency of use of analytical and narrow-bore HPLC columns, then the narrow-bore HPLC columns are highly popular [8]. One reason is the relative sensitivity at the detector, which is enhanced by about a factor of 10 in comparison with that attained by analytical columns (Table 3), another reason may be the typical flow rates of 50 to 500 µL/min, which are very compatible with some of the commonly employed LC/MS interfaces.

Although the use of analytical and narrow-bore HPLC columns reflects the current state of the art, the future belongs to capillary liquid chromatography and miniaturised LC/MS systems. A decisive reason for this development is the excellent sensitivity of this HPLC technique. Thus, for example, "on-column" detection limits have been recorded at the low femtomole level for the determination of peptides from cell culture supernatant liquid [18]. New developments apply a high direct current (20 to 50 kV) over the entire length of the micro-bore and nano-bore column to diminish a possible band broadening and for an additional enhancement of the separation power from a maximum of 40 000 to 65 000 theoretical plates in the original capillary liquid chromatography [16] to 200 000 theoretical plates [8, 20]. The increase in separation power is attributed to an improvement in the mass interaction between the stationary and the liquid phases. The new analytical technique is used in two types of application: capillary electrochromatography (CEC) and pseudo-capillary electrochromatography (pCEC). Like capillary electrophoresis, the liquid phase is transported solely by electro-osmosis in CEC [20]. As in classical HPLC, additional flow is generated by pressure in the case of pCEC [21, 22].

A greater availability of miniaturised HPLC systems, a broad range of commercially available high-quality micro-bore and nano-bore columns at reasonable prices and the development of robust methods, especially LC/MS procedures to detect low analyte concentrations in complex matrices – in particular with regard to analysis in biological materials – will be of decisive importance for the future success of the different variations of capillary liquid chromatography in routine laboratory practice.

3.1.2.2 Sample preparation for routine analysis (multi-dimensional HPLC)

The use of LC/MS methods often permits sample preparation to be reduced to a minimum, as urine and blood samples can be analysed as aqueous solutions without further derivatisation after separation of the protein components by precipitation using the HPLC solvents methanol or acetonitrile and subsequent centrifugation [23–25]. Thus losses of the analyte due to sample processing are reduced and sensitivity is simultaneously improved. However, matrix effects may sometimes have adverse effects that influence the chromatographic separation (e.g. broadening of the signals). In addition, a matrix-rich analytical solution may reduce (but rarely enhance) the ion yield. In most cases it leads to more rapid build-up of contamination in the complete system. As a rule, the consequences are lower sensitivity as well as diminished robustness of the method. This leads to shorter maintenance intervals in order to restore the capability to perform sensitive measurements. Thus the time saved by reducing sample preparation and the advantage of higher sensitivity due to lower losses of the analyte may be nullified [26–29]. In a review Mallet et al. describe a method of quantifying ion suppression or ion yield enhancement and point out some ways of avoiding these effects [30]. In addition, the authors explain pH effects and their consequences on the ion yield of various analytes and on ionisation polarity.

If sample preparation is extremely minimised, samples should be filtered and the HPLC columns should be protected from matrix contamination by guard columns.

Blood and urine contain many salts and organic compounds that are readily soluble in water; these compounds are generally eluted when the proportion of organic solvent is low, and they therefore suppress the ionisation yield of an analyte that is eluted under the same conditions. By selecting the HPLC conditions (type of column, gradient elution) so that the analyte leaves the column with as high a proportion of organic eluent as possible, improvement in sensitivity is obtained; in addition it is possible to use a switch valve to divert the eluate that does not contain the analyte to a waste container. This diminishes contamination of the mass spectrometer.

Further procedures have been developed to increase the sensitivity and degree of automation of LC/MS analyses. Zell et al. published a method for analysing larger volumes without overloading the analytical HPLC column and contaminating the entire analysis system and thus nullifying the improved sensitivity [19, 31, 32]. This system, now known as column-switching, carries out a shortened form of column chromatography to achieve enrichment and pre-cleaning of the sample with the aid of a 6-port valve and a "trap" column. For this purpose up to several hundred microlitres of test solution are loaded onto the trap column together with an eluent containing as high a proportion of water as possible. More lipophilic analytes are bound on the column, whereas polar components are washed into a waste vessel. After the washing procedure the analytes are eluted from the trap column onto the analytical column with the aid of a gradient pump and an eluent with a higher organic proportion. Then the actual chromatographic separation can be performed as described above. Applications of this principle for analyses in biological materials can be found in Koch et al., Brink et al. and Kellert et al. [33–35]. As several millilitres of sample material are normally available for analysis, column-switching represents a convenient method of attaining very low detection limits.

The transfer of the target analytes from the trap column to the analytical column can be carried out in the same direction as the flow onto the trap column or the direction of flow can be reversed in the so-called "back-flush" procedure (see Figure 2). As a rule, the back-flush technique leads to transfer of the target analytes onto the analyti-

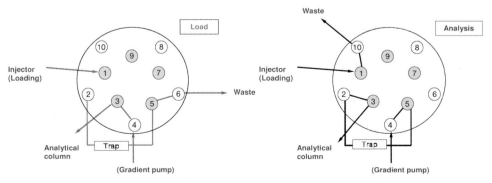

Fig. 2. Schematic illustration of a column-switching system

cal column as sharply defined bands, with consequential advantages for chromatographic separation and quantitative determination.
Other variations of column-switching using only one column (in this case the trap column is used for sample preparation and serves as the analytical column too) as well as the use of 3 columns (with two trap columns in parallel or two analytical columns in parallel) are described in reviews of this subject [36–38].

3.1.2.3 Trap columns

Trap columns that permit direct injection of material containing protein without previous protein precipitation have now been specially developed for analyses in biological materials (urine, blood, plasma, sputum). Depending on the analytical investigation, only hydrolysis (to cleave phase II conjugates) and/or pH adjustment have to be carried out before injection.
Separation of the protein matrix by these biocompatible trap columns, which have become commercially available only in recent years, is achieved by means of various physical separation principles. Reviews of this field [36, 38, 39] provide comprehensive overviews. The trap columns mentioned below all have the advantage that, depending on the demands of the analysis, up to several mL of urine or up to 100 µL of serum/plasma can be introduced in one injection without chromatographic problems. Depending on their type and the application, the useful life of these trap columns is given as several hundred cycles, and this is also attained. Centrifugation of the samples before injection prolongs the useful life of trap columns. The duration of a chromatographic separation run depends on the type of trap column used, but a fully automated run is prolonged only by between 0.5 and at most 10 minutes and considerable time is saved in sample preparation.

Restricted access media (RAM)
The term "restricted access media" (RAM) describes a family of column materials that permits direct injection of biological fluids, in which only low-molecular substances (<15 to 20 kDa) can interact with the active groups of the column material (RP materials with C4, C8, C18, nitrile or phenyl groups). Conversely, high-molecular compounds are denied access to active sites on account of their size (pore size or chemical barriers). In addition, a hydrophilic coating on the outer surface of the column material prevents adsorption of the protein matrix [39, 40].
RAM phases that exclude the protein matrix due to the specific pore size of the basic material are commercially available in different variations. ISRP ("Internal Surface Reversed Phase") columns (from Regis Technologies) are inaccessible to molecules with a size greater than 20 kDa and retain hydrophobic molecules on a tripeptide phase (glycine-L-phenylalanine-L-phenylalanine) in the internal pores (e.g. GFF and GFF II) mainly by means of interaction with the π electrons [41, 42]. These columns are supplied in the dimensions of both pure trap columns (1 cm×3 mm) and analytical columns (up to 25 cm×4.6 mm). ISRP columns have proved especially efficient in the direct analysis of serum [37, 40] but applications in the analysis of

urine have also been reported [42]. ADS (alkyl-diol-silica) phases (Merck) with a pore size of 6 nm exclude molecules with a size greater than approx. 15 kDa. Internal coatings with C4, C8 and C18 reverse phases are currently obtainable [39]. These ADS columns have been specially developed as trap columns (2.5 cm×4 mm) and are at present the most popular RAM phases for biological fluids. The wide range of applications, both in serum and in urine as well as in other biological fluids (sputum, microdialysate, milk, etc.), is described comprehensively in the literature [36, 38, 40]. The analyses of the phthalate metabolites in urine [35] and serum [43] give currently valid examples (see also "Di(2-ethylhexyl)phthalate (DEHP) metabolites" in this volume). The use of the ADS phase has been limited in particular by the fact that only C4, C8 and C18 classical reverse phases have been commercially available as internal coatings to date, and retention of even moderately polar target analytes is poor. The ChromSpher 5 Biomatrix column (Chrompack/Varian) with a pore size of 13 nm retains small molecules by hydrophobic interactions with phenyl groups. However, the reported range of application of this column is rather narrow [38].

A second type of RAM phase achieves the exclusion of proteins and other macromolecules mainly or additionally by means of a chemical barrier. In SPS ("semipermeable surface") phases (Regis Technologies) proteins are excluded by a semipermeable polyoxyethylene coating of the silica material, while retention occurs in the underlying nitrile, phenyl, C4 or C18 phases that are accessible only to smaller molecules. Like ADS columns, SPS columns are remarkable for their robustness, whereby SPS columns offer a wider range of reverse phase retention [44]. As in the case of GFF columns, SPS columns are available in the dimensions of both pure trap columns (1 cm×3 mm) and analytical columns (up to 25 cm×4.6 mm). The basic silica material (C8 and C18) or the hydrophobic basic polymer material of BioTrap and BioTrap MS columns (Chromtech) is coated with the human plasma protein a_1-acid glycoprotein (AGP). The BioTrap MS column can be used in a relatively wide pH range from 2 to 10. BioTrap columns are reported to be especially suitable for analysis in serum and plasma on account of the coating with AGP [38, 45]. Capcell PAK columns (Shiseido/Phenomenex) consist of a silica material with a silicone monolayer coating, whereby active silanol groups are protected by alkyl chains. Polyoxyethylene chains prevent the adsorption of macromolecules. Depending of the type of Capcell, retention of molecules of less than 20 kDa is achieved by interaction with C1, C8, C18, phenyl, CN, NH_2 or cation-exchanger (SCX) groups. Eight different types of Capcell columns are now commercially available, some of which are also suitable for retention of relatively polar compounds (e.g. MG-II, AQ types). In addition, the Capcell columns are remarkable for their applicability over a wide pH range (1 to 10). Column dimensions range from purely trap columns (1 cm×4 mm) to analytical columns (up to 25 cm×4.6 mm). Several publications confirm the wide-ranging applicability of Capcell columns for the analysis of urine, serum and plasma samples in one-dimensional and multi-dimensional separation systems [38, 46–49]. Another type of column, which is specially marketed for direct injection of serum samples, is Supelcosil Hisep (Sigma-Aldrich). This silica-based material is protected from protein bonding by polyethylene oxide, while low-molecular compounds are bound by phenyl groups [38]. Column dimensions from 2 cm×4.0 mm to 25 cm × 4.6 mm are available.

Large particle support (LPS)
Another approach that enables direct injection of biological fluids is the use of column materials with relatively large particle diameters (25 to 60 µm). "Turbulent flow chromatography" (TFC) makes use of columns with small inner diameters (between 0.5 and 2.1 mm) that are operated at high flow rates (3 to 5 mL/min). This changes the laminar flow profile into a turbulent flow profile. Analytes with a molecular size of <1000 Da can be retained in the hollow spaces in this way. Conversely, macromolecules and salt clusters are not retained. TFC leads to a very good exchange between the stationary and the mobile phase with the result that the concentrated analyte can be effectively transferred from this trap column to the analytical column. This technique makes further sample pre-treatment unnecessary, especially in the case of matrices containing protein (plasma, tissue homogenates). Detailed descriptions can be found in [50–59]. On principle, two types of column material are used in this application: (1) conventional silica material with e.g. C2, C8, C18 or phenyl coating and (2) polymer material (to date divinylbenzene-N-vinylpyrrolidone copolymer). In general, all the column materials known in "off-line" SPE can also be used in TFC. Both types of column permit several hundred plasma injections of 100 µL without recognisable deterioration in the analytical performance. This applies only if the sample is introduced under turbulent flow conditions. Under laminar flow conditions protein accumulation soon renders the column unusable [60]. It is also remarkable that, on account of the high flow rates, introduction and washing cycles of less than 1 min are sufficient before the sample is transferred to the analytical column and analytical separation is achieved by means of column-switching and by a flow rate reduction to a range of 0.25 to 0.5 mL/min.

Oasis columns (Waters) have been developed specially for the requirements of TFC and are now offered in various modifications and in column dimensions from 2 cm×2.1 mm up to 2 cm×4.6 mm. Oasis HLB (hydrophilic lipophilic balance) is especially suitable for the retention of lipophilic molecules by divinylbenzene groups, whereas moderately polar to hydrophilic molecules are retained by N-vinylpyrrolidone groups. Thus Oasis HLB covers a much broader analytical spectrum than conventional C18 materials. Other Oasis modifications, such as Oasis MCX, Oasis WAX, Oasis MAX and Oasis WCX, combine the RP phase properties of HLB with anion and cation exchanger characteristics of varying strength. Therefore at present Oasis columns provide the widest range of retention mechanisms for on-line sample preparation with trap columns. Numerous publications describe the use of Oasis columns for on-line extraction of analytes from biological fluids [38, 61–66].

A second LPS technique that is also quite new is the use of monolithic columns for direct injection of biological fluids. In this case too (like column material with large particle diameters) sample injection can be carried out at high flow rates without generating high pressure on account of the high permeability of these columns. The flow rates are 5 to 10 times higher than in the conventional application of monolithic columns. Under these conditions the higher-molecular protein matrix is also eluted with the solvent front. Applications for plasma as well as urine samples have been described [38, 67, 68]. At present, the Chromolith Flash RP-18e column (Merck) seems especially suitable for use as a trap column (dimensions of 2.5 cm×4,6 mm) [69].

Further on-line techniques
Some further on-line techniques that have also been made commercially available recently must be mentioned in this context. On-line solid phase extraction (SPE) that can be integrated into the software and hardware of the common LC/MS systems appears promising, as it permits fully automated analysis of plasma and urine samples without further processing. In this case the SPE columns are automatically replaced by the instrument after being used once or several times, depending on wear and tear. This system generally serves to separate undesirable matrix components rather than to enrich the analyte. The first publications indicate that this advance is promising and readers should keep an eye on future developments [70].
When very hydrophilic compounds (glucuronide and sulphate metabolites) are analysed with a higher proportion of organic eluent, it may prove impossible to elute the analyte by HPLC separation with conventional C18 materials. An isocratic chromatographic separation must be performed with a very low percentage of organic solvent (<5%), e.g. to prevent elution of the analyte with the dead volume of the column. Acetonitrile or other organic solvents that form a maximum azeotropic mixture with water can be added to the eluate through a T-junction after chromatographic separation of the analyte. This facilitates the transition of the analyte molecules into the gaseous phase, and despite dilution due to addition of the solvent, the ion yield is enhanced as shown by the example of ethylglucuronide [71]. It is also possible to increase the ion yield by reducing the HPLC flow rate at the end of the separation column. For this purpose the eluate can be divided into portions after the HPLC column and a smaller portion can be transferred to the LC/MS interface. Although much less analyte reaches the mass spectrometer, this loss is more than compensated by the improved ion yield. We refer readers to the study carried out by Gangle et al. for further details [72].

3.1.3 Capillary electrophoresis (CE)

Capillary electrophoresis (CE) is a micro-separation technique for organic compounds that are readily ionised in solution. It is based on application of a high direct current (15 to 30 kV) to the entire length of a "fused-silica" capillary column (inner diameter: ≤100 µm) filled with buffer solution. This leads to electro-osmotic flow of the solution at a rate of 0 to 200 nL/min with a drop-like profile. The different migration rates of the ions in the solution under the influence of the electric field can result in a very effective separation of the analyte(s) from the matrix. Technically five separation principles (modes) are employed: capillary zone electrophoresis (CZE), isotachophoresis (ITP), micellar electrokinetic chromatography (MEKC), capillary gel electrophoresis (CGE) and capillary isoelectrofocusing (CIEF) [73, 74].

The theoretical advantages of this separation technique are minimisation of sample volumes, low consumption of solvents, short separation times, high resolution and simple method development. At present, however, CE is still in the transition phase from research to routine practice. Potential clinical users are hoping for the development of instruments that are completely suitable for routine use and that permit a high sample throughput as well as for ready-to-use applications in the form of user-friendly kits containing all the necessary components [73].

With regard to LC/MS, efforts have been made to use CE as a separation system since the 1980s [75–77].

The main problem in coupling CE with a mass spectrometer is posed by the low flow rates. They are typically in the range of <1 to 100 nL/min. In contrast, the most frequently used interfaces are operated at flow rates ranging from 1 to 200 µL/min. Various measures can be taken to compensate for this difference: on the one hand the flow rate at the end of the capillary can be increased by feeding external liquid ("sheath-flow" CE interfaces and "liquid junction" CE interfaces), on the other hand hopes are pinned on a general miniaturisation of interfaces ("sheathless-flow" CE interfaces) [20]. As described in Section 3.1.1 the infeed of liquid leads to dilution of the analyte and to a possible loss in sensitivity. But miniaturisation of the interfaces requires considerable technical ingenuity and users need special training to handle these LC/MS systems, so neither of the two approaches has yet born fruit in efforts to make this technique suitable for routine analysis.

The low maximum loading capacity of fused silica capillary columns places another obstacle in the path towards routine use of CE, especially with regard to analyses in biological materials. As described for capillary liquid chromatography (Section 3.1.2) columns can easily become overloaded, particularly when the analyte(s) is/are present in complex matrices. In this case too, an attempt can be made to use "on-column" enrichment techniques so that samples with volumes at the µL level with low analyte concentrations can be introduced [76, 78].

In their review article of 1995 Cai and Henion [75] discussed the low concentration-dependent sensitivity of the CE/MS coupling ($\geq 10^{-5}$ M) that hindered the use of CE/MS for routine analysis at that time. However, in their opinion the technique would be improved in the thereafter, resulting in wider distribution among users. In 2002 Gelpí [20] came to the conclusion that these expectations had not yet been fulfilled. Thus CE/MS coupling has still not become a routine analytical technique – also with regard to analysis in biological materials – but remains an interesting field of research.

3.2 LC/MS interfaces

As already discussed above, the interface plays a central role in LC/MS coupling. In the preceding (almost) forty years different technical solutions have been offered to overcome the problem of transforming analyte molecules or ions present in the liquid phase to analyte ions in the gaseous phase. Depending on whether the analyte molecule is first enriched and then ionised (sampling/ionisation) or first ionised and then enriched (ionisation/sampling), the interfaces employed for on-line LC/MS procedures can be assigned to one of two categories [20]. The first category comprises the "moving wire/moving belt" interfaces, the "particle beam" interface, the "direct liquid introduction" interface and the "continuous flow fast atom bombardment" interface. The second category consists of atmospheric pressure ionisation (API) sources, such as electrospray ionisation (ESI), the "atmospheric pressure chemical ionisation" (APCI) source and the "atmospheric pressure photo-ionisation" (APPI) interface as well as its precursor the thermospray interface.

3.2.1 Classical LC/MS interfaces

The interfaces mentioned in the first category and the thermospray interface belong to the classical LC/MS interfaces. These classical LC/MS interfaces have greatly declined in importance for routine analysis in recent years since the introduction of robust commercially available atmospheric pressure ionisation sources. They are still suitable for discrete applications in some cases, but manufacturers of some interface types have discontinued their production, e.g. thermospray interfaces [5]. For these reasons classical LC/MS interfaces will not be discussed in detail in this section. Interested readers are referred to publications on the basic studies and reviews given in Table 4.

3.2.2 Atmospheric pressure ion sources

It is characteristic for atmospheric pressure ion sources that analyte molecules are ionised at atmospheric pressure and then subsequently enriched and transferred to a mass spectrometer. Three methods of ionisation are possible for routine analysis and they form the physico-chemical basis for the design of commercially available instrumental systems: the application of high voltage electric fields (ESI), electrical discharge (APCI) and irradiation with a UV source (APPI). Whereas the number of publications describing classical LC/MS interfaces for analysis has decreased in re-

Table 4. Literature overview of classical LC/MS interfaces

LC/MS interface	Basic studies	Reviews
Moving wire/moving belt	[79, 80]	[81, 82]
Particle beam	[83, 84]	[85, 86]
Direct liquid introduction	[2–4, 87–89]	[90, 91]
Continuous flow fast atom bombardment	[92, 93]	[94, 95]
Thermospray	[96–99]	[100–102]

cent decades, the number of articles on the use of ESI and APCI sources has risen sharply, with studies on the APPI interface reinforcing this trend since 2000 [6, 103]. The three types of LC/MS interface will be presented in detail below, the fundamental physico-chemical processes will be introduced, and the schematic instrumental set-up (with variations if applicable) will also be explained.

3.2.2.1 Electrospray ionisation (ESI)

The first applications of ESI were reported by Fenn et al. [104–106] and by Aleksandrov et al. [107] independently of each other in the mid-1980s. Figure 3 shows the electrical circuit on which ESI is based.

A sharply pointed steel capillary (needle) serves as one electrode, and a collector (aperture plate) functions as the counter-electrode. A high potential of several kilovolts is applied between the two electrodes, whereby oxidation occurs at the needle and reduction at the collector. The eluent from the LC completes the electrical circuit. A prerequisite for detection by means of ESI-MS is that the analyte must be present as an ion in the eluent. The applied electric field ensures a partial separation of the positively and negatively charged ions in the eluent by partly penetrating the liquid surface at the tip of the steel capillary. In the positive ion mode shown here, positively charged ions are enriched at the tip of the needle, while negatively charged ions are forced back into the interior of the steel needle. In the negative ion mode the poles are reversed and the contrary effect is achieved.

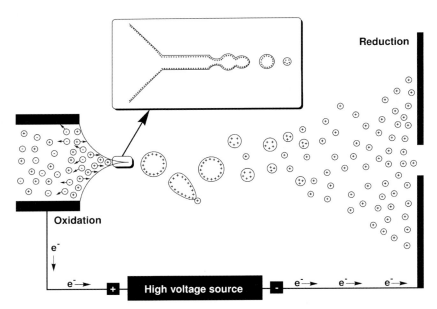

Fig. 3. The electrospray process (according to [108])

The surface tension of the liquid is overcome due to the mutual repulsion at the surface and the simultaneous attraction of the positively charged ions to the counter-electrode under the influence of the electric field; a cone of liquid forms (the Taylor cone) with its tip being the last stable point that joins a thin thread of liquid, which in turn breaks down to individual charged droplets (see Figure 3, enlargement). The following equation gives the electric field strength E_0 required to initiate the electrospray process:

$$E_0 = \left(\frac{2\gamma \cos 49°}{\varepsilon_0 r_c}\right)^{1/2}$$

whereby γ is the surface tension of the liquid, cos 49° is half the angle of the Taylor cone, ε_0 stands for the electric field constant and r_c is the radius of the steel capillary. The surface tension of the liquid has a remarkable influence on the process. Liquids such as water have such a high surface tension that the vacuum field strength required for the electrospray process can lead to electrical corona discharges, especially in the negative ion mode. Special precautions must be taken to suppress these undesirable discharges when ESI is used [109]. A stable electrospray that forms a Taylor cone requires LC flow rates of 0.5 to 5 µL/min [6] and ion concentrations of $\geq 10^{-5}$ up to $< 10^{-3}$ M at the tip of the needle, with ion concentrations of approx. 10^{-5} M representing optimum conditions for charge separation [108].

A counter stream of hot nitrogen (typical temperatures: 200 to 300 °C, typical gas flow: 1 to 12 L/min) evaporates liquid from the tiny charged droplets formed by the spray process on their path from the needle to the collector. This diminishes the droplet size and the surface charge density increases. If the surface charge density exceeds the critical value of 10^8 V/cm^3 (the "Rayleigh limit") the drops are subjected to tear-shaped deformation due to the repulsive forces of like charges, and a stream of fragments is emitted [110, 111]. The resulting total surface of the fragments is higher than that of the original droplet. If it is assumed that the initially present charges have a statistically uniform distribution on the resulting fragments, then the surface charge density is reduced and falls below the critical Rayleigh limit again.

Depending of the size of the original drop, the released fragments may be smaller droplets, which are further evaporated and in turn emit fragments when the Rayleigh limit is exceeded anew. This process proceeds until no more liquid can be evaporated and the ions either crystallise as solids or pass into the gaseous phase. This description of the electrospray process is known in the literature as the "charge residue model" (CRM) [1, 112, 113]. It is also possible that ions are directly emitted from the liquid phase into the vapour phase from droplets with a radius of 10 to 20 nm. The surface charge density required for this is below the Rayleigh limit. This is known as the "ion evaporation model" (IEM) [114, 115].

Which model or which combinations of both models best explains the electrospray process is the subject of current scientific discussion. However, IEM seems to better reflect the behaviour of smaller molecular ions during ESI, whereas CRM provides a very good description of the physico-chemical processes in the case of larger molecular ions, e.g. globular proteins [108].

ESI is notable for three characteristics. Firstly, it is a very mild ionisation technique that is highly efficient in generating protonated (or deprotonated molecular ions in the case of the negative ion mode) from polar, even thermally labile, compounds with a high molecular weight. Secondly, it generates multiply charged ions of the [M + nH]$^{n+}$ type (analogue ions are observed in the negative ion mode) from molecules with multiple basic groups. Thirdly, the molecular ions formed directly reflect the acid-base equilibrium in the LC eluent because they are formed by protonation/deprotonation of the molecules in solution or by formation of adducts with solvent ions. Therefore ESI is especially suitable for the detection of compounds that can be readily ionised in solution [6].

The formation of multiply charged ions has particularly wide-reaching consequences for the use of ESI. Mass spectrometers measure mass-to-charge ratios (m/z). Signals of molecular ions bearing multiple charges appear in a lower mass range than the singly charged molecular ion. Thus, for example, a protein with a 5-fold charge and a molecular weight of 8 kD yields measurable signals at 1600 m/z (m: 8000/z: 5). Whereas the signal of a singly charged molecular ion would not be detectable at a measurement range of 50 to 2000 m/z by the mass spectrometer, the signals of the protein bearing a 5-fold charge can be readily detected after ESI. As a rule, however, a molecular ion with a 5-fold charge does not generate one single signal, but a series of signals from differently charged ions emanating from the same molecule ([M + nH]$^{n+}$; n: 1, 2, 3, 4, 5, 6, ...). The real molecular weight of the analyte can be calculated from the distribution and intensity of the observed signals with the aid of special computer programs known as "deconvolution" software. Thus a quadrupole mass spectrometer is capable of determining the molecular weights of compounds up to 50 kDa with a mass deviation of less than 0.01% [116].

Figure 4 shows the schematic set-up of an electrospray interface ("curtain gas" type). The electrospray process comprehensively described above takes place between the needle and the aperture plates. An additional heated glass capillary installed between the aperture plate and the ion optics often aids focusing and final desolvation of the generated ions. The ion optics (skimmer, octopole, lenses) focus the ion beam before it enters the mass spectrometer. The high vacuum necessary for the mass spectrometer is achieved by several vacuum chambers arranged in series. The pump technology used to create the vacuum has already been described in the General Introduction to Volume 10 of this series.

The linear arrangement of the needle and the aperture plates shown here permits optimal entry of the ions generated by the electrospray process and thus enables the greatest possible sensitivity, but at the same time it poses the risk of clogging due to deposits of liquid(s) or solid(s) from the spray. Various commercial suppliers have attempted to overcome this problem by alternative arrangements of the needle and aperture plates or by special inlet systems [6, 117]. However, according to our current knowledge only the orthogonal arrangement (90° angle) of the needle seems to permit a comparable degree of sensitivity to that achieved by the linear arrangement while simultaneously enhancing the robustness of the system.

As already mentioned, a stable electrospray is created at LC flow rates of 0.5 to 5 µL/min; but the typical flow rates of narrow-bore or analytical columns currently

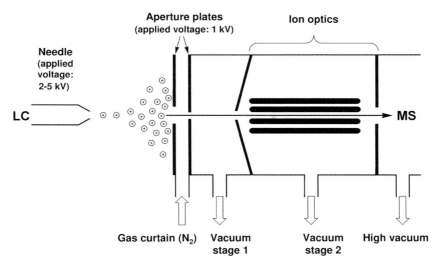

Fig. 4. Schematic set-up of an electrospray interface ("curtain gas" type described in [5])

used in routine analysis are higher by factors of 10 to 200. Therefore aerosol formation is optimised in the commercially available electrospray interfaces in order to establish the necessary compatibility. This can be achieved by the use of a nebuliser gas coaxially introduced into the needle ("ionspray": flow rate < 1.0 mL/min [118]), by ultrasound ("ultraspray": flow rate < 0.5 mL/min [119]) or by heat ("turbospray": flow rate > 0.1 mL/min [120]). As discussed in Sections 3.1.2 and 3.1.3 the use of miniaturised LC/MS systems in routine analysis seems to offer a promising option for the future, even for analyses in biological materials. Special electrospray interfaces for low flow rates (< 200 nL/min to 4 µL/min), known as "micro ESI" and "nanospray", that are currently still undergoing development and testing will play an important role in these advances [8, 20].

3.2.2.2 Atmospheric pressure chemical ionisation (APCI)

Modern atmospheric pressure chemical ionisation (APCI) sources are based on the ground-breaking developments pioneered by Horning et al. in the mid-1970s [121, 122]. Figure 5 shows the schematic set-up of a currently used APCI interface ("tube" type).
The liquid components of the eluent and the compounds dissolved in it are suddenly but still gently evaporated with the aid of a nebuliser and auxiliary gas introduced into the LC eluent through a coaxial capillary and using heat (300 to 500 °C) [5, 117]. Corona discharges from a discharge needle under high voltage (5 to 10 kV) generate free electrons. Alternatively, a ^{63}Ni beta source provides electrons. The emitted electrons can interact with the gases present in the ionisation zone, such as

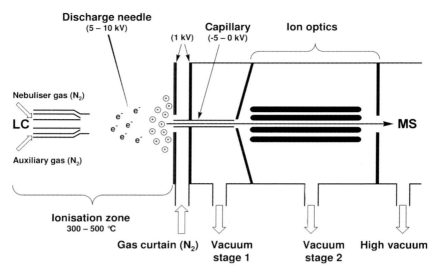

Fig. 5. Schematic set-up of an APCI interface ("tube" type as described in [5, 117])

nitrogen (N_2), oxygen (O_2), nitrogen monoxide (NO) and water (H_2O). This process creates a plasma that is rich in various reactant ions, the most important component being the solvated hydronium ion $H_3O^+(H_2O)_n$ (Figure 6).
In the positive ion mode the reactant ions in the plasma, especially the solvated hydronium ion $H_3O^+(H_2O)_n$, transfer protons or positive charges to neutral analyte molecules in the gaseous phase, whereas in the negative ion mode ionisation is achieved by proton removal or electron capture [124]. The proton affinity of water vapour (165 kcal/mol) decisively influences whether ionisation of the analyte molecule in the gaseous phase takes place or not, as solvated hydronium ions $H_3O^+(H_2O)_n$ make up the main proportion of the reactant ions in the positive ion mode [125–127]. Only compounds with a higher proton affinity than water vapour become and remain protonated.
A detailed description of the mechanisms of chemical ionisation together with lists of the proton affinities of various compounds are found in the General Introduction to Volume 10.
The APCI ionisation process is extremely efficient and under ideal conditions the ion yield can be 100%. However, it should be noted that the analyte molecules compete with the matrix components of the sample and with the solvent molecules of the eluent for the reactant ions in the plasma. This can considerably influence the detection sensitivity of the relevant analyte(s) [5, 125–127]. The mass spectra that result from APCI generally exhibit a very prominent molecular ion of the compound to be detected in addition to a few fragment ions [122].
The ions formed are transmitted through a cone or a heated glass or metal capillary into the mass spectrometer in the commercially available instruments. As in ESI, the nebuliser has a linear or orthogonal arrangement with respect to the inlet to the mass

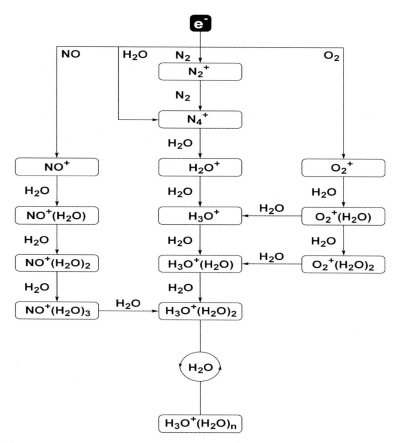

Fig. 6. The formation of the solvated hydronium ion $H_3O^+(H_2O)_n$ at the APCI interface according to [5, 123]

spectrometer; in this case too, desolvation of the ions formed by interaction with the plasma ions is often achieved with the help of a counter-stream of hot nitrogen. The ion optics used (skimmer, octopole, lenses) and the arrangement of several vacuum chambers in series to attain the required high vacuum are identical to the technology used in ESI. This permits manufacturers to offer users LC/MS instruments in which an ESI source can be exchanged for an APCI interface, depending on the desired application. In such cases the same basic instrument can be equipped to perform analysis in both ionisation modes.

The typical LC flow rates that are compatible with APCI sources are between 0.5 and 2.0 mL/min. Such sources are thus ideal for use with analytical columns [5].

APCI has proved successful for the detection of ionisable polar and non-polar molecules with a molecular weight <2000 u or in cases where the LC eluents are relatively non-polar solvents [5, 6]. In particular, it is widely applied in the pharmaceutical industry. APCI is an excellent complement to the analytical range accessible to ESI, which does not cover the above-mentioned applications. In other cases, namely the determination of biopolymers, APCI is inferior to ESI. For the above reasons and, as the technical changeover is simply performed, the complementary use of both ionisation techniques brings important advantages to many analytical laboratories.

3.2.2.3 Atmospheric pressure photo-ionisation (APPI)

Atmospheric pressure photo-ionisation (APPI) is a very new ionisation method for LC/MS. Robb et al. reported the first APPI source in 2000 [128]. Starting with a commercial APCI interface, they replaced the discharge needle by a vacuum ultraviolet lamp that emitted 10 eV photons. As in the case of the APCI source, the liquid eluent containing the dissolved compounds is evaporated suddenly but gently by a combination of a nebuliser and auxiliary gas (introduced coaxially with respect to the LC eluent) and by heat (450 °C) before the generated gases are irradiated by photons in the immediately adjacent photo-ionisation zone. There the following reaction can proceed in the positive ion mode:

$$M + h\nu \longrightarrow M^{\bullet+} + e^-$$

This involves the absorption of a photon by a molecule in the gaseous phase, the release of an electron and the formation of a positively charged molecule radical ion. This always occurs when the incident photon energy is greater than the ionisation energy of the molecule. Especially when protic solvents, elevated source pressure and prolonged photon irradiation are used, the protonated molecular ion is also found in addition to the dominant molecule radical ion [129].
Despite this, there is a low statistical probability that an analyte molecule will be ionised. To enhance the yield of analyte ions a suitable, readily photo-ionisable substance (dopant) can be introduced into the stream of auxiliary gas through a T-junction. After being initially photo-ionised, the dopant transfers its charge directly to the analyte molecule (reaction 1) if the electron affinity of the analyte is greater than that of the dopant. However, if the proton affinity of the analyte molecule is greater than that of the deprotonated dopant, solvent molecules and their clusters serve as intermediates for proton transfer from the dopant ion to the analyte molecule (reaction 2) [103, 130]:

$$D + h\nu \longrightarrow D^{\bullet+} + e^-$$
$$1)\ D^{\bullet+} + A \longrightarrow D + A^{\bullet+}$$
$$2)\ D^{\bullet+} + L \longrightarrow [D-H]^{\bullet} + LH^+$$
$$LH^+ + A \longrightarrow L + AH^+$$

Anisole, benzene, toluene and acetone are commonly used dopants; the possible use of phenol is being investigated at present [103]. Acetonitrile represents a special case among the LC solvents, as its solvent clusters can be directly photo-ionised and they can subsequently ionise analyte molecules [131, 132].

In the negative ion mode mainly deprotonated analyte ions (accompanied by adducts with solvent molecules and fragmentation reactions in the source) are formed from acidic analyte molecules in the gaseous phase by means of electron capture, dissociative electron capture or ion-molecule reactions, or analyte anions (accompanied by substitution products) are generated from analyte molecules with a positive electron affinity [103] (see General Introduction to Volume 10).

To date two commercial suppliers offer instruments with APPI interfaces. As in the case of ESI and APCI one manufacturer prefers a linear, the other an orthogonal arrangement of the nebuliser with respect to the inlet to the mass spectrometer. The set-up of the mass spectrometer and the inlet is identical to the known elements of instruments with other API sources.

In numerous applications narrow-bore columns are recommended and lower flow rates (e.g. 0.2 mL/min) than those recommended for the APCI source are suggested for the coupling of APPI interfaces with LC.

The spectrum of the substances that can be detected with APPI is largely identical to that already described for APCI on account of the great similarity in the ionisation processes. However, it has been tentatively concluded that detection limits lower than those yielded by other API sources may be achieved for more labile compounds, such as steroids, quinones or amino and nitro aromatic substances, due to the milder conditions in the APPI interface; in particular compared with APCI [103, 133, 134]. It remains to be seen whether this trend will be confirmed in the future; if this is indeed the case the value of APPI as an ionisation method will be further enhanced.

3.3 Mass filters

One or several mass filters can be coupled with the LC/MS interface. These filters separate the molecular ions according to their mass-to-charge ratios. The physicochemical principles of mass filters are comprehensively described in the General Introduction to Volume 10.

3.3.1 Mass filters in general

In general, the mass filters used for GC/MS are also widely employed in liquid chromatography/mass spectrometry. These include magnetic sector field [135], quadrupole [136] and ion trap instruments [137]. The individual instruments and their advantages and drawbacks have been comprehensively presented in the "General Introduction" to Volume 10. In addition, this chapter includes a detailed discussion of the principles of tandem mass spectrometry (MS/MS) using the triple quadrupole mass spectrometer and the ion trap as examples. Therefore we will discuss only mass filters that are used in LC/MS and mass filter combinations and that were not previously presented in the DFG "Analyses of Hazardous Substances in Biological Materials" collection of methods below.

3.3.2 Special mass filters

Additional mass filters that are employed in LC/MS are the time-of-flight (TOF) mass spectrometer, Fourier transform mass spectrometer (FT-MS) and the orbitrap. TOF instruments, especially in combination with other mass filters, are now well established in routine analysis; in contrast, the FT mass spectrometer and the orbitrap are mainly used in basic analytical research.

3.3.2.1 Time-of-flight mass spectrometer

Time-of-flight mass spectrometers (TOF-MS) are based on a suggestion put forward by Stephens in the 1940s [138] and studies by Wiley and McLaren [139] in the 1950s. Figure 7 shows the principle of the time-of-flight mass spectrometer.

Molecular ions leave the ion source (e.g. an API source) and reach a zone where they are accelerated by an electric field. Then they travel a defined distance in the flight tube at a speed that depends on their mass before reaching the detector where their arrival is recorded. The time between the application of the acceleration impulse to the ions and their arrival is measured. The following equation applies:

$$t = \left(\frac{2md}{eE}\right)^{1/2} + L\left(\frac{m}{2eV_0}\right)^{1/2}$$

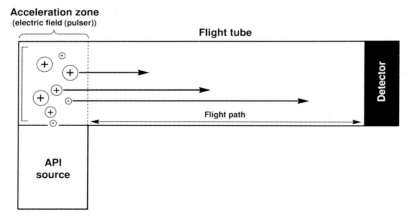

Fig. 7. Principle of the time-of-flight mass spectrometer (modified according to [5])

whereby t designates the flight time, m the molecular mass, d the length of the acceleration zone, e the ion charge, E the strength of the applied electrostatic field, L the length of the flight path without a field and V_0 the applied acceleration voltage. Typical flight times of molecular ions range from 5 to 100 microseconds [140].

Whereas general mass filters record the spatial separation of the ions to yield mass spectra, temporal separation of the ions in the TOF mass spectrometer has two decisive advantages: the (theoretically) unlimited mass range and the extremely rapid and sensitive recording of complete mass spectra [5, 141–143]. In this case all the incoming ions are simultaneously registered, so the time-consuming scanning of a previously defined mass range becomes unnecessary. Thus $>10^6$ u/s can be recorded compared with 4000 u/s for quadrupole and ion trap instruments [5]. However "selected ion monitoring" (SIM) for further enhancement of sensitivity is impossible due to instrumental design.

One disadvantage of a linear time-of-flight mass spectrometer (Figure 7) is its low resolution power R = 1000 [140]. This is due to the difference in the time the ions of identical mass required for formation in the source region, to their different spatial arrangement in the electric field of the acceleration zone, and to the variation in the initial kinetic energy of the ions before they receive the acceleration impulse in the direction of the flight tube [5]. Technical improvements in the acceleration of the ions [139], and in particular the introduction of the reflectron by Mamyrin et al. [144], have considerably increased the resolution (R = 8000 to 10 000 [141]) in modern time-of-flight mass spectrometers (Figure 8).

The electric fields of the reflectron reduce the speed of the ions and reflect them in the direction of the detector. If ions with the same mass-to-charge ratio but with different kinetic energy reach the reflectron, then the ions with the higher energy (and

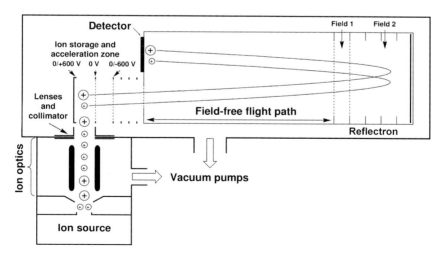

Fig. 8. Schematic set-up of a modern time-of-flight mass spectrometer with orthogonal injection and reflectron (modified according to [142])

velocity) penetrate deeper into the electric fields of the reflectron than the ions with lower energy (and velocity). Thus the former spend more time in the reflectron and their velocity is adjusted to that of the slower ions with the same mass-to-charge ratio on their further flight path. This ensures that all the ions with the same mass-to-charge ratio reach the detector simultaneously.

The coupling of an API source to a mass spectrometer that is typical for modern LC/MS proves more difficult in the case of the TOF mass spectrometer than for general mass filters (Section 3.3.1). The reason is that API sources, such as the ESI interface, generate a continuous ion beam that can be readily detected by continually operating mass spectrometers such as quadrupole spectrometers. In contrast, discontinuous acceleration impulses applied to the ions are absolutely necessary for measurements in TOF mass spectrometry. An orthogonal arrangement of the API source with respect to the TOF mass spectrometer creates an ion storage and acceleration zone from which discrete "packets" of the ion beam that is being continuously generated by the API source can be accelerated into the flight tube (Figure 8). Ion optics installed before the flight tube permit the velocity of the ions to be minimised by collisional damping and cause focusing of the ion beam, which leads to heightened sensitivity and resolution [141, 143]. In this way complete mass spectra of the eluted components can be obtained, even from narrow peaks (elution time: 0.1 to 1 s) like those generated by micro-bore columns [143].

The TOF mass spectrometer has proved successful with LC/MS coupling on account of its (theoretically) unlimited mass range and the extremely rapid and sensitive recording of the complete mass spectra of peptides and especially of proteins with high molecular weights. In tandem mass spectrometry it often replaces the third quadrupole and allows the highly sensitive recording of complete mass spectra of the daughter ions (Section 3.3.3).

3.3.2.2 Fourier transform mass spectrometer

The fundamental principle behind Fourier transform mass spectrometers (FT mass spectrometers) or ion cyclotron resonance mass spectrometers (ICR mass spectrometers) was described in the 1930s by E.O. Lawrence [145], and it was first applied in mass spectrometers by Marshall et al. in 1974 [146].

FT mass spectrometers are also known as "magnetic ion traps". The ions, e.g. from an API source, reach the trap where they oscillate around the axis of a very strong magnetic field that is applied. Their frequency of rotation is inversely proportional to their mass. The rotation of the ions induces an electrical current from which their frequency of rotation can be determined. The measured frequency of rotation is converted to mass by Fourier transformation [5].

The advantages of FT mass spectrometry are the large mass range (>15 000 u) that can be measured and the extremely high resolution (<500 000). Moreover, in contrast to all other mass filters, the ions are not destroyed during detection, which permits highly sensitive determination of product ions following multiple dissociation reactions (MS^n experiments). One great drawback is the high cost involved in purchasing and maintaining the instruments. Therefore at present FT mass spectrometers are mainly used in basic analytical research for structural clarification of compounds. In a recent review [147] the author points out that numerous commercial suppliers are currently endeavouring to introduce "benchtop" FT mass spectrometers onto the market. It remains to be seen whether these instruments will also provide a feasible option in the near future for laboratories performing routine analysis where they may be coupled with LC mainly for structural clarification.

3.3.2.3 Orbitrap

The most recent development of the ion trap mass spectrometer is the "orbitrap" [148]. This ion trap contains a central, spindle-shaped electrode. The ions are injected into the orbitrap tangentially with respect to this electrode, and they orbit around the central electrode as a result of electrostatic attraction. At the same time they oscillate back and forth along the axis of the central electrode. The frequency of these oscillations generates electric signals at the detector plate, which are converted to the corresponding m/z ratios by Fourier transformation. The principle is therefore similar to ICR-MS, but it functions with an electrostatic field instead of a magnetic field. For this reason orbitraps do not require a complex cooling system with liquid helium, as no supraconducting magnetic field has to be generated. The mass resolution power of orbitraps is only slightly poorer than an ICR-MS instrument with a 7-Tesla magnet. We refer interested readers to the primary literature for more detailed information [149–151]. As these instruments are considerably less expensive to maintain than FT mass spectrometers, orbitraps coupled with LC will certainly be useful for both structural clarification of proteins and for the detection of "smaller" molecules in the future.

3.3.3 Combinations

Tandem mass spectrometry has already been introduced in Volume 10 with the triple quadrupole and ion trap serving as examples. The significance of different scanning techniques in LC/MS will be briefly discussed below using well-known examples.
Triple quadrupole instruments are equipped with various scanning techniques that can be applied individually or in combination. The "selected reaction monitoring" (SRM) or the "multiple reaction monitoring" (MRM) mode is especially suitable for the necessary sensitive and exact quantification of analyses in biological materials. A characteristic precursor ion is selected in a SIM (selected ion monitoring) measurement procedure in the first quadrupole (Q1). Then it is accelerated against a collision gas by a certain voltage in the second quadrupole (Q2), which functions as a collision cell, and the resulting product ion is selectively detected in the third quadrupole (Q3). Such SRM or MRM transitions are extremely characteristic and are accompanied by a low background signal. Therefore it is possible to achieve very specific and sensitive detection of a compound.

The electronic parameters can be rapidly changed in the quadrupoles and the collision cell can be quickly evacuated again due to the special quadrupole geometry. In this way contamination due to successive analytes (known as "cross talk") can be prevented and a large number of target ions can be detected simultaneously and quantitatively evaluated.

A further scanning technique is the "product ion" scan that can yield important information for structural clarification. Like the MRM technique, it is based on the collision of precursor ions with collision gases.

A function that also serves for structural clarification is the "neutral loss" scan, which is based on the principle of fragmentation of a precursor ion. However, in this case the cleavage of a neutral fragment that cannot be detected directly is measured indirectly. This technique is important for the determination of glucuronides, sulphates or mercapturic acids, which cleaved as neutral particles from the xenobiotic parts. Thus these metabolites can be readily identified and quantified in complex matrices. Measurement is based on the principle that Q1 and Q3 carry out scans staggered by the difference in the mass of the neutral particle, and only those mass transitions that are exactly equivalent to this difference in mass are detected, e.g. m/z = 176 u in the case of a glucuronide. The "precursor ion" scan is the 4th mode that may be important for structural analysis. In this case a product ion is permanently measured in Q3, while Q1 simultaneously scans a pre-defined mass range. Whenever the product ion is detectable, it is accurately assigned to a precursor ion and this information can be used for structural clarification. For example, modified DNA bases can be found in this way, as unmodified bases form a stable product ion as a rule.

An MS/MS coupling known as "tandem in space" is required for the "neutral loss" and "precursor ion" scanning modes, as the MS/MS analysis steps must be spatially separated (Q1, Q2 = collision cell, Q3), e.g. in the triple quadrupole instrument. A "product ion" scan can also be performed by an ion trap.

In recent years further combinations have been introduced by commercial suppliers for use in routine analysis to complement the conventional tandem mass spectrometers. These include the triple quadrupole with the linear ion trap (Qtrap) and a combination of two quadrupoles with a TOF mass spectrometer (QqTOF).

3.3.3.1 Combination of a triple quadrupole and an ion trap (Qtrap)

Qtrap technology is a combination of a tandem mass spectrometer (triple quadrupole mass spectrometer) and a linear ion trap, whereby the second analytical quadrupole (Q3) can serve as a quadrupole or as a linear ion trap. Such a mass spectrometer incorporates the technology to perform all the types of scans that are possible using a traditional tandem mass spectrometer and can also be operated with the second analytical quadrupole acting as a linear ion trap, e.g. to perform MS^n experiments.

However, "in space" fragmentation, typical for tandem mass spectrometers, is carried out in the MS^2, whereas the MS^3, like the ion trap, operates on the principle of "in time" fragmentation. "In space" fragmentation as a result of collisions of the selected precursor ions with inert gas molecules is more effective than "in time" fragmentation in ion traps. For this reason the software of such combinations usually restricts the MS^n experiments to MS^3. Besides more effective cleavage, "in space" fragmentation has the additional advantage that all fragments are detected, which is physically impossible in an ion trap. The product ion spectra thus obtained show the typical fragmentation patterns of a tandem mass spectrometer.

This instrument offers several additional advantages for analyses in biological materials. Compared with traditional tandem mass spectrometers or ion traps considerably more sensitive product ion spectra can be recorded in the "enhanced product ion" (EPI) mode. This offers the possibility of identifying unknown substances by the acquisition of complete product ion spectra. In addition, these fragments can be fragmented again and analysed by MS^n for further characterisation.

An intelligent software solution, known as "information-dependent acquisition", allows different types of scanning to be freely combined during the same analysis, e.g. a "constant neutral loss" (CNL) scan followed by an EPI scan. For instance, glucuronic acids result in a CNL of m/z = 176 u. If this CNL is combined with an EPI, a corresponding EPI spectrum is recorded for each detected CNL signal. For example, intensive CNL signals alone may be selected by means of freely adjustable parameters, then a mass spectrum can be recorded with EPI, or mass spectra with different collision energies can also be generated to provide additional information for characterisation. This coupling permits more information to be acquired in a shorter time with comparable sensitivity. There is no necessity for multiple measurements.

The combination of an MRM scan with an EPI scan yields very high sensitivity and highly reliable identification, as the corresponding product ion spectrum is available for each MRM signal.

Thus, for example, the combination of an MRM with an EPI scan improves the reliability of identification in the detection of glucuronides by means of MRM, in which the transition of M$^{+/-}$ to [M$^{+/-}$ -176 u] is the most sensitive transition in many cases. In this case a complete mass spectrum replaces an additional qualifier transition [24]. As urine contains a large number of glucuronides, so that other isobar glucuronides with this characteristic transition are observed as well as the target analyte, this combination of scans permits reliable detection of the glucuronide in question.

Reviews on the application of the Qtrap technique are to be found in [152–154].

3.3.3.2 Combination of quadrupole and TOF (QqTOF)

QqTOF represents a further hybrid instrument that integrates a TOF mass spectrometer for detection instead of the second analytical quadrupole. In addition to ion traps, QqTOF instruments are frequently used in the mass spectrometric investigation of proteins to characterise peptides. In the "small" molecule range it is suitable for characterising unknown analytes and metabolites due to its high mass accuracy of <10 ppm. The high mass accuracy of the QqTOF instruments permits exact determination of the empirical formulae of unknown analytes. The structural formula can be established with the aid of appropriate software and of databases. Therefore laboratories in the chemical industry investigating metabolism are the main users, as QqTOF instruments enable structural clarification of unknown metabolites of new active substances in the pharmaceutical and pesticide sectors. Comparison of QqTOF instruments with other LC/MS devices and a comprehensive discussion on the advantages and disadvantages of this technique are found in [155–157].

3.4 Detectors

The detectors commonly employed in LC/MS (secondary electron multiplier or photon multiplier) are the same as those used for GC/MS. A detailed description of their function is given in the General Introduction to Volume 10.

4 Aspects of the development and optimisation of methods

4.1 General aspects

As described in detail in Section 3.2, the ionisation process in LC/MS takes place at atmospheric pressure using the normal ESI, APCI and APPI sources. Figure 9 shows the application ranges of each type of source.

The criteria for the selection of a suitable mass spectrometer or a combination of mass spectrometers are presented in the General Introduction to Volume 10 (sector field device, quadrupole instrument, ion trap) or in Section 3.3 (TOF mass spectrometer, FT mass spectrometer, orbitrap) of this chapter.

The following substance groups are possible candidates for detection in biological materials with the aid of LC/MS. All the metabolites of phase II metabolism, such as sulphates, glucuronides, acetates, amino acid conjugates and the mercapturic acids

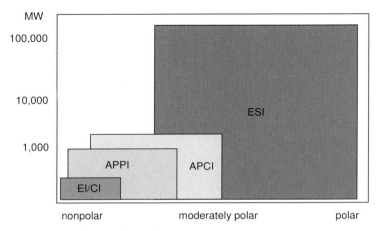

Fig. 9. Operational ranges of the individual ion sources

from glutathione-S conjugates are basically suitable for detection and quantification by means of LC/MS. In many cases DNA adducts, organic acids and compounds containing nitrogen can also be sensitively analysed as biomarkers by LC/MS. Conversely, aldehydes, ketones or very non-polar substances are rather unsuitable. However, methods have been described for LC/MS analysis of aldehydes, but they require prior derivatisation [158, 159]. Table 5 contains a selection of current LC/MS methods for the assay of xenobiotics in biological materials. As several hundred articles describing LC/MS applications were published in 2006 alone, this table can only present a selective overview. It specifies the substance itself and the metabolite class to which it belongs, as biological monitoring uses the metabolite as a biomarker in addition to the parent compound in some cases. As a rule the xenobiotics/metabolites are hydrophilic analytes that are suitable for detection by HPLC and that can be analysed with widely used ionisation techniques such as ESI and/or APCI. Thus, for example, bisphenol A can be very sensitively determined both with the electrospray [24] and with APCI ionisation techniques [160]. Blood, plasma, serum and urine samples are generally analysed in the field of biological monitoring. In some cases samples of human milk are investigated to detect xenobiotics. Independent of these different sample matrices, the same LC/MS method can be applied after appropriate sample processing. In some cases protein precipitation may be the only preparation step necessary for all matrices, e.g. for urine and human milk [160, 161].

It is no simple matter to transfer methods from one laboratory to another or from one LC/MS instrument to another on account of the different ionisation techniques (especially ESI and APCI) and due to differences in the individual interfaces from different manufacturers. In contrast to GC/MS, generally applicable spectra libraries are not available. They are only valid for the manufacturer concerned and generally only for a specific type of instrument. Nevertheless, the molecular ions and mass fragments of an analyte generated by different instruments are comparable, but they often differ in their intensities or their ratios. This also applies to molecular ions, as M+X cluster ions may also be generated, depending on the interface and ionisation parameters. Further information must therefore be taken from the primary literature, as many parameters that cannot be discussed in detail here have to be adjusted to adapt a method.

In addition to calibration of the instrument with appropriate standard substances (reserpine, polypropylene glycol mixtures) that are generally specified by the manufacturer, the entire process of ionisation, ion separation and detection is carried out specifically for the relevant analyte. This fine adjustment is known as "tuning". This is decisively different to the procedure for GC/MS, as the calibration and tuning are usually performed with a tuning substance independent of the analyte. Therefore tuning is a relatively time-consuming process in LC/MS analysis, especially when several compounds are to be analysed, but it is absolutely essential in order to achieve sensitive measurements.

The influence of the LC eluent and its composition as well as the influence of buffer solutions or the addition of acid on the ionisation process can be tested and then optimised to obtain the highest sensitivity. If available, different ion sources can be compared with regard to the highest attainable ion yield for the relevant substance.

Table 5. Examples of LC/MS methods to detect xenobiotics in biological materials

Xenobiotic	References
Mercapturic acids	
Acrylamide	[34, 162–164]
Benzene	[165]
1-Bromopropane	[166]
1,3-Butadiene	[167]
Styrene	[168]
Vinyl chloride	[169, 170]
Xylene	[171]
Glucuronides	
Aromatic amines	[172]
Bisphenol A	[24, 160, 173]
1-Bromopropane	[166]
Daidzein	[174]
Ethanol	[71, 175–177]
n-Hexane	[178, 179]
Nicotine	[180]
PAH	[181]
Pesticides	[182]
Phthalates	[43, 183–187]
Tetrabromobisphenol A	[188]
Sulphates	
1-Bromopropane	[166]
Ethanol	[71, 175–177]
Pesticides	[182]
PAH	[181]
Organic acids	
Benzene	[165]
Methyl parathion	[189]
Organophosphates	[190]
Trichloroethene and tetrachloroethene	[191]
Parabens	[192, 193]
Perfluorinated surfactants	[194–196]
Phthalates	[184, 197–199]
Aromatic amines	
Aminobiphenyl derivatives	[200]
Heterocyclic aromatic amines	[201, 202]
Nitro aromatic compounds	[133, 134]
DNA adducts	
Dimethylnitrosoamine	[33]
Oxidative stress	[33, 203]

As a rule, quantitative analyses are carried out at LC flow rates in the range of 150 to 400 µL/min (in the case of APCI up to 1 mL/min). This leads to an ionisation yield that is far less constant than that achieved by GC/MS. Ionisation is also affected by impurities in the LC eluent and in the inert gases used to assist evaporation of the eluent as well as contamination of the source by matrix components that are not removed by the vacuum system at atmospheric pressure. Although reduced sample preparation and coupling with an autosampler permit several hundred analyses to be performed per day, thus saving time and lowering the cost of analysis, serious contamination of the ion source results when complex biological matrices such as blood and urine are analysed. Compared with past generations, however, the latest instruments are comparatively robust under these adverse conditions.

In general, the use of internal standards for quantitative analysis by means of LC/MS considerably increases the accuracy of the method. Analytes labelled with isotopes (^{2}H, ^{13}C, ^{15}N or ^{18}O) are to be recommended as internal standards insofar as they are available on the market or can be readily synthesised. As the chemical and physical behaviour of these compounds differs only minimally from that of the actual analyte, they are eminently suitable for the compensation of errors or interferences (e.g. during sample preparation or in the ion source).

4.2 Development and optimisation of methods

Figure 10 shows the general procedure to be followed for the development and optimisation of an LC/MS method.

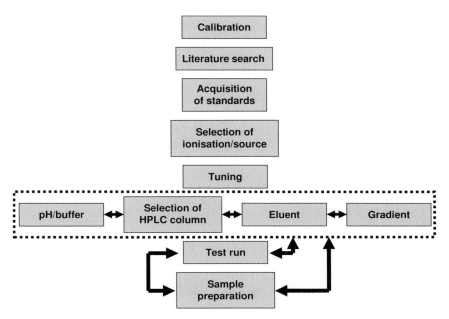

Fig. 10. The individual steps to be followed for the development and optimisation of an LC/MS analytical method

Calibration

The instrument should be initially calibrated with the standard substances (e.g. polyethylene glycol standard) specified by the relevant manufacturer in order to ensure that the exact mass-to-charge ratio m/z is obtained during tuning. The procedure to be followed is given in the instructions provided by the commercial supplier of the mass spectrometer.

Literature search

A search for relevant publications should be the first step in method development in order to save time. Important information is obtained when a search targets the ionisation technique (ESI, APCI, APPI), the charge (positive, negative), the LC conditions used (column, capillary, buffer, organic solvent, gradient, flow rate) for the substance or class of substance concerned, e.g. under "Materials and Methods" in publications or in collections of methods such as the one presented here. Data on the detection limit and the validity criteria provide additional indications of the applicability and the reliability of the method. In rare cases, for example, if the method is described for the same analytical equipment (LC and MS system) and the given detection limit is satisfactory for the desired application, the method can be taken over in its entirety and tuning becomes unnecessary.

Tuning

If no published data are available, the structural formula of the analyte provides the first important indications. Lipophilic substances, such as steroids, nitro aromatic compounds, etc., are more readily ionised by APPI or APCI than by ESI. It should also be considered whether the analyses are not easier to perform by means of GC/MS. In contrast, hydrophilic analytes, such as all phase II metabolites, are generally very suitable for the ESI technique. Substances with acidic protons (organic acids, phenols, etc.) are more readily analysed in the negative ionisation mode, whereas compounds that can easily accept a proton (many compounds containing nitrogen) are better detected using the positive ionisation mode.
Numerous standard substances are commercially available today. However, some substances (e.g. certain phase II metabolites) have to be synthesised in-house or by a contract laboratory. The glucuronides represent a special group of standards, as their synthesis proves very laborious. However, they can be isolated in small quantities by microsomal incubation or from the urine of experimental animals or from human urine [204, 205]. As no weighable amounts are generally obtained in this way, the standard solution of glucuronide can be quantified either by UV spectroscopy or following enzymatic cleavage of the starting substance. Calibration solutions can then be prepared from this solution [206].

In the first experimental step the analyte or the standard substance is dissolved in the solvents normally used for LC/MS (water, acetonitrile, methanol) and dilutions in the concentration range from 1 to 100 ng/μL are prepared. If the positive ionisation mode is to be used, the ion yield can be considerably enhanced by acidification of the solution to a pH of 1 using formic acid. In contrast, acidification tends to diminish the ion yield in the negative ionisation mode. Then the MS parameters for the substance under investigation are "tuned" with the aid of the syringe pump (Section 3.1.1). Depending on the type of instrument, flow rates ranging from 5 to 50 μL/min are necessary for this procedure when ESI is used. APCI and APPI often require higher flow rates, and APPI also requires the addition of dopants such as toluene (Section 3.2.2.3).

In addition, solvents that are filled into the syringe pump must be degassed and free from bubbles in order to achieve a stable spray and thus a stable signal.

Initially, the default gas flow and voltage settings of the relevant mass spectrometer are used in the "full scan" mode. The mass range must cover the molecular ion mass of the analyte. If the relevant $[M+H]^+$ or $[M-H]^-$ ion of the substance is detected with a signal-to-background noise ratio (S/N) of more than 10 under these conditions, then the optimisation process can be initiated by adjusting the instrumental voltages. Depending on the manufacturer, the relevant instrumental software may offer "autotuning" functions that save the user the task of manually optimising the different voltages. Such a function saves considerable time, particularly in the case of triple quadrupole instruments, as more than 5 different voltages must be optimised, against each other in some cases. More details are found in the appropriate user manuals.

Integration/optimisation of the LC

The next step is to integrate the liquid chromatographic system, usually an HPLC system. The chromatographic behaviour of the analyte is influenced by the choice of a suitable HPLC separation column (Section 3.1.2) and HPLC solvents, by the use of buffer solutions (whereby the use of more volatile salts, such as ammonium formate is preferable to less volatile salts), and by suitable gradients. These parameters are selected to ensure that as high an S/N ratio as possible is attained at the highest possible proportion of organic solvent in the HPLC eluate and at low flow rates of 150 to 300 μL/min (in the case of ESI sources). More modern ESI sources operate at higher flow rates and are optimised for these flow rates. APCI and APPI sources also run at higher flow rates. We refer readers to the relevant user manuals.

An HPLC-UV detector connected in series can also assist this optimisation. A high proportion of organic solvent and slow flow rates result in a very "dry" electrospray, i.e. the eluent components can be evaporated very effectively and an optimum ion yield is achieved.

The final step is to optimise the gas flow rates under the established HPLC conditions and, if necessary, the voltages to be applied, especially in the ion source, to attain the best possible ion yield. In the case of a single quadrupole instrument the mass spectrometer is operated in the "selected ion monitoring" (SIM) mode. In contrast, tandem mass spectrometry (e.g. ion trap or triple quadrupole instrument) must be operated in the "selected reaction monitoring" (SRM) or "multiple reaction monitoring" (MRM) mode. Two procedures are followed:

A. No HPLC column is used, but the flow rate is set at the value that has been established for chromatographic separation. The solvent ratio of the HPLC eluent is that at which the analyte is eluted from the HPLC column. Identical quantities of an analyte standard mixture are injected under these conditions and the gas flow and instrumental voltages are adjusted until the highest signal-to-background noise ratio is obtained.

B. The HPLC column to be used can be connected or not, but the flow rate is set at the value that has been established for chromatographic separation. The HPLC solvent composition is that at which the analyte is eluted from the HPLC column. In this procedure the analyte standard mixture already used for tuning is mixed with the eluate from the HPLC column through a T-junction, e.g. by means of a syringe pump (Section 3.1.1) and at a flow rate of 5 to 50 µL/min. This procedure permits continual optimisation of the appropriate parameters (which saves time as a rule) and is advisable for the tuning process with an APCI or APPI ion source, as these operate at higher flow rates.

In both cases it is advisable to prepare the analyte standard mixture in the eluent mixture in which the analyte is eluted from the HPLC column. This prevents the occurrence of any double signals that may occur (after the liquid chromatographic system has been integrated) due to the fact that the analyte is not fully retained on the separation column as a result of an excessive proportion of organic solvent. If the organic proportion is too low, the analyte may not be completely dissolved in some cases.

The first analytical run can then be started with the complete system and all the parameters can be checked again. The process described above must be carried out anew for other analytes. A separate tuning process is not generally required in the case of isotope-labelled internal standards. However, when a triple quadrupole instrument is used, it may be necessary to check the fragmentation of the internal standard, especially if the product ions cannot be clearly assigned. If no isotope-labelled standards are available, a standard substance must be selected that most closely resembles the analyte to be determined with regard to its ionisation and chromatographic behaviour. The complete tuning process must be carried out to ensure that this condition is fulfilled.

The necessary indices for evaluation of the method (precision, accuracy, detection limit and interference) are determined on the basis of calibration solutions. If possible, these should be prepared in the matrix that will later become the sample matrix.

5 Perspectives

Liquid chromatography/mass spectrometry (LC/MS) combines the liquid chromatography (LC) and mass spectrometry (MS) analytical techniques. LC/MS instruments have been commercially available from various suppliers for the past two decades.

Most of the earlier publications on LC/MS coupling emanated from pharmacokinetic laboratories of the pharmaceutical industry. This technology has almost completely superseded GC/MS in this field, as hydrophilic metabolites of medicinal substances are ideal analytes for the LC-ESI-MS technique. Therefore it is not surprising that this technique is becoming increasingly established in biological monitoring. For instance, in the year 2000 there were 67 entries under the term "Monitoring AND LC/MS" in the "Pub-Med" database, in 2003 the number had grown to 133 and in 2006 it had reached 251.

In view of this rapid development it is not easy to speculate on the future of LC/MS. But various trends are recognisable, some of which have already been pointed out at appropriate points in the text and are briefly mentioned here again. Further miniaturisation of the systems in the field of liquid chromatography and accompanying enhancement of sensitivity at the detector are to be expected. This also applies to the development of LC/MS interfaces; miniaturisation is also probable in this case. Advances in sample preparation will minimise manual activities and progress towards complete automation (e.g. automated on-line SPE on special column material). The introduction of "benchtop" FT mass spectrometers and wider use of the orbitrap is foreseeable. At the same time this opens up opportunities for new mass spectrometer combinations. Therefore, with a view to biological monitoring, further improvement of detection sensitivity and specificity may be expected from future developments.

6 References

[1] *M. Dole, L.L. Mack, R.L. Hines, R.C. Mobley, L.D. Ferguson* and *M.B. Alice:* Molecular beams of macroions. J. Chem. Phys. 49, 2240–2249 (1968).

[2] *V.L. Tal'roze, G.V. Karpov, I.G. Gorodetskii* and *V.E. Skurat:* Capillary system for introducing liquid mixtures into an analytical mass spectrometer. Zh. Fiz. Khim. 43, 3104–3112 (1968).

[3] *V.L. Tal'roze, G.V. Karpov, I.G. Gorodetskii* and *V.E. Skurat:* Analysis of mixtures of organic substances in a mass spectrometer with a capillary system for introducing liquid samples. Zh. Fiz. Khim. 43, 367–372 (1969).

[4] *V.L. Tal'roze, V.E. Skurat* and *G.V. Karpov:* Use of a mass spectrometer with a capillary system for admitting the sample as a liquid chromatograph detector. Zh. Fiz. Khim. 43, 452–453 (1969).

[5] *R. Willoughby, E. Sheehan* and *S. Mitrovich:* A global view of LC/MS – How to solve your most challenging analytical problems. 2nd Edition, Global View Publishing, Pittsburgh, Pennsylvania, USA (2002).

[6] *J. Abian:* The coupling of gas and liquid chromatography with mass spectrometry. J. Mass Spectrom. 34, 157–168 (1999).

[7] *G.D. Bowers, C.P. Clegg, S.C. Hughes, A.J. Harker* and *S. Lambert:* Automated SPE and tandem MS without HPLC columns for quantifying drugs at the picogram level. LC-GC 15, 48–53 (1997).

[8] *J. Abian, A.J. Oosterkamp* and *E. Gelpí:* Comparison of conventional, narrow-bore and capillary liquid chromatography/mass spectrometry for electrospray ionization mass spectrometry: practical considerations. J. Mass Spectrom. 34, 244–254 (1999).

[9] *K.B. Tomer, M.A. Moseley, L.J. Deterding* and *C.E. Parker:* Capillary liquid chromatography/mass spectrometry. Mass Spectrom. Rev. 13, 431–457 (1994).

[10] *J.P. Chervet, M. Ursem* and *J.P. Salzmann:* Instrumental requirements for nanoscale liquid chromatography. Anal. Chem. 68, 1507–1512 (1996).

[11] *J.P.C. Vissers, H.A. Claessens* and *C.A. Cramers:* Microcolumn liquid chromatography: instrumentation, detection and applications. J. Chromatogr. A 779, 1–28 (1997).

[12] G. Crescentini, F. Bruner, F. Mangani and Y. Guan: Preparation and evaluation of dry-packed capillary columns for high-performance liquid chromatography. Anal. Chem. 60, 1659–1662 (1988).
[13] M.T. Davis and T.D. Lee: Analysis of peptide mixtures by capillary high performance liquid chromatography: A practical guide to small-scale separations. Protein Sci. 1, 935–944 (1992).
[14] J.C. Gluckman, A. Hirose, V.L. McGuffin and M. Novotny: Performance evaluation of slurry-packed capillary columns for liquid chromatography. Chromatographia 17, 303–309 (1983).
[15] S. Hoffman and L. Blomberg: Packed capillary columns for liquid chromatography. Chromatographia 24, 416–420 (1987).
[16] R.T. Kennedy and J.W. Jorgenson: Preparation and evaluation of packed capillary liquid chromatography columns with inner diameters from 20 to 50 micrometers. Anal. Chem. 61, 1128–1135 (1989).
[17] T. Takeuchi and D. Ishii: High-performance micro packed flexible columns in liquid chromatography. J. Chromatogr. A 213, 25–32 (1981).
[18] A.J. Oosterkamp, E. Gelpí and J. Abian: Quantitative peptide bioanalysis using column-switching nano liquid chromatography/mass spectrometry. J. Mass Spectrom. 33, 976–983 (1998).
[19] M. Zell, C. Husser and G. Hopfgartner: Low picogram determination of Ro 48-6791 and its major metabolite, Ro 48-6792, in plasma with column-switching microbore high-performance liquid chromatography coupled to ion spray tandem mass spectrometry. Rapid Commun. Mass Spectrom. 11, 1107–1114 (1997).
[20] E. Gelpí: Interfaces for coupled liquid-phase separation/mass spectrometry techniques. An update on recent developments. J. Mass Spectrom. 37, 241–253 (2002).
[21] T. Eimer, K.K. Unger and J. van der Greef: Selectivity tuning in pressurized-flow electrochromatography. Trends Anal. Chem. 15, 463–468 (1996).
[22] E.R. Verheij, U.R. Tjaden, W.M.A. Niessen and J. van der Greef: Pseudo-electrochromatography-mass spectrometry: a new alternative. J. Chromatogr. A 554, 339–349 (1991).
[23] W. Völkel, R. Alvarez-Sanchez, I. Weick, A. Mally, W. Dekant and A. Pähler: Glutathione conjugates of 4-hydroxy-2(E)-nonenal as biomarkers of hepatic oxidative stress-induced lipid peroxidation in rats. Free Radic. Biol. Med. 38, 1526–1536 (2005).
[24] W. Völkel, N. Bittner and W. Dekant: Quantitation of bisphenol A and bisphenol A glucuronide in biological samples by high performance liquid chromatography-tandem mass spectrometry. Drug Metab. Dispos. 33, 1748–1757 (2005).
[25] S. Wagner, K. Scholz, M. Donegan, L. Burton, J. Wingate and W. Völkel: Metabonomics and biomarker discovery: LC-MS metabolic profiling and constant neutral loss scanning combined with multivariate data analysis for mercapturic acid analysis. Anal. Chem. 78, 1296–1305 (2006).
[26] H.C. Bi, G.P. Zhong, S. Zhou, X. Chen and M. Huang: Determination of adefovir in human plasma by liquid chromatography/tandem mass spectrometry: application to a pharmacokinetic study. Rapid Commun. Mass Spectrom. 19, 2911–2917 (2005).
[27] I. Ferrer, J.F. Garcia-Reyes, M. Mezcua, E.M. Thurman and A.R. Fernanndez-Alba: Multi-residue pesticide analysis in fruits and vegetables by liquid chromatography-time-of-flight mass spectrometry. J. Chromatogr. A 1082, 81–90 (2005).
[28] H.R. Liang, R.L. Foltz, M. Meng and P. Bennett: Ionization enhancement in atmospheric pressure chemical ionization and suppression in electrospray ionization between target drugs and stable-isotope-labeled internal standards in quantitative liquid chromatography/tandem mass spectrometry. Rapid Commun. Mass Spectrom. 17, 2815–2821 (2003).
[29] H. Mei, Y. Hsieh, C. Nardo, X. Xu, S. Wang, K. Ng and W.A. Korfmacher: Investigation of matrix effects in bioanalytical high-performance liquid chromatography/tandem mass spectrometric assays: application to drug discovery. Rapid Commun. Mass Spectrom. 17, 97–103 (2003).
[30] C.R. Mallet, Z. Lu and J.R. Mazzeo: A study of ion suppression effects in electrospray ionization from mobile phase additives and solid-phase extracts. Rapid Commun. Mass Spectrom. 18, 49–58 (2004).
[31] M. Zell, C. Husser, R. Erdin and G. Hopfgartner: Simultaneous determination of a potassium channel opener and its metabolite in rat plasma with column-switching liquid chromatography

using atmospheric pressure chemical ionisation. J. Chromatogr. B Biomed. Sci. Appl. 694, 135–143 (1997).

[32] *M. Zell, C. Husser* and *G. Hopfgartner:* Column-switching high-performance liquid chromatography combined with ionspray tandem mass spectrometry for the simultaneous determination of the platelet inhibitor Ro 44-3888 and its pro-drug and precursor metabolite in plasma. J. Mass Spectrom. 32, 23–32 (1997).

[33] *A. Brink, U. Lutz, W. Völkel* and *W.K. Lutz:* Simultaneous determination of O6-methyl-2'-deoxyguanosine, 8-oxo-7,8-dihydro-2'-deoxyguanosine, and 1,N6-etheno-2'-deoxyadenosine in DNA using on-line sample preparation by HPLC column switching coupled to ESI-MS/MS. J. Chromatogr. B Analyt. Technol. Biomed. Life Sci. 830, 255–261 (2006).

[34] *M. Kellert, K. Scholz, S. Wagner, W. Dekant* and *W. Völkel:* Quantitation of mercapturic acids from acrylamide and glycidamide in human urine using a column switching tool with two trap columns and electrospray tandem mass spectrometry. J. Chromatogr. A 1131(1–2), 58–66 (2006).

[35] *H.M. Koch, L.M. Gonzalez-Reche* and *J. Angerer:* On-line clean-up by multidimensional liquid chromatography-electrospray ionization tandem mass spectrometry for high throughput quantification of primary and secondary phthalate metabolites in human urine. J. Chromatogr. B Analyt. Technol. Biomed. Life Sci. 784, 169–182 (2003).

[36] *M.J. Berna, B.L. Ackermann* and *A.T. Murphy:* High-throughput chromatographic approaches to liquid chromatographic/tandem mass spectrometric bioanalysis to support drug discovery and development. Anal. Chim. Acta 509, 1–9 (2004).

[37] *D. Ortelli, S. Rudaz, S. Souverain* and *J.-L. Veuthey:* Restricted access materials for fast analysis of methadone in serum with liquid chromatography-mass spectrometry. J. Sep. Sci. 25, 222–228 (2002).

[38] *S. Souverain, S. Rudaz* and *J.L. Veuthey:* Restricted access materials and large particle supports for on-line sample preparation: an attractive approach for biological fluids analysis. J. Chromatogr. B Analyt. Technol. Biomed. Life Sci. 801, 141–156 (2004).

[39] *K.S. Boos* and *C. Grimm:* High-performance liquid chromatography integrated solid-phase extraction in bioanalysis using restricted access precolumn packings. Trends in Anal. Chem. 18, 175–180 (1999).

[40] *E.A. Hogendoorn, P. van Zoonen, A. Polettini, G. Marrubini Bouland* and *M. Montagna:* The potential of restricted access media columns as applied in coupled-column LC/LC-TSP/MS/MS for the high-speed determination of target compounds in serum. Application to the direct trace analysis of salbutamol and clenbuterol. Anal. Chem. 70, 1362–1368 (1998).

[41] *V.K. Boppana, C. Miller-Stein* and *W.H. Schaefer:* Direct plasma liquid chromatographic-tandem mass spectrometric analysis of granisetron and its 7-hydroxy metabolite utilizing internal surface reversed-phase guard columns and automated column switching devices. J. Chromatogr. B Biomed. Appl. 678, 227–236 (1996).

[42] *J. Martínez Fernández, J.L. Martínez Vidal, P. Parrilla Vázquez* and *A. Garrido Frenich:* Application of restricted-access media column in coupled-column RPLC with UV detection and electrospray mass spectrometry for determination of azole pesticides in urine. Chromatographia 53, 503–509 (2001).

[43] *H.M. Koch, H.M. Bolt* and *J. Angerer:* Di(2-ethylhexyl)phthalate (DEHP) metabolites in human urine and serum after a single oral dose of deuterium-labelled DEHP. Arch. Toxicol. 78, 123–130 (2004).

[44] *Z. Yu, D. Westerlund* and *K.S. Boos:* Evaluation of liquid chromatographic behavior of restricted-access media precolumns in the course of direct injection of large volumes of plasma samples in column-switching systems. J. Chromatogr. B Biomed. Sci. Appl. 704, 53–62 (1997).

[45] *A. El Mahjoub* and *C. Staub:* High-performance liquid chromatographic method for the determination of benzodiazepines in plasma or serum using the column-switching technique. J. Chromatogr. B Biomed. Sci. Appl. 742, 381–390 (2000).

[46] *M. Katagi, M. Nishikawa, M. Tatsuno, A. Miki* and *H. Tsuchihashi:* Column-switching high-performance liquid chromatography-electrospray ionization mass spectrometry for identification of heroin metabolites in human urine. J. Chromatogr. B Biomed. Sci. Appl. 751, 177–185 (2001).

[47] *H.M. Lee, S.J. Choi, C.K. Jeong, Y.S. Kim, K.C. Lee* and *H.S. Lee:* Microbore high-performance liquid chromatographic determination of cisapride in rat serum samples using column switching. J. Chromatogr. B Biomed. Sci. Appl. 727, 213–217 (1999).

[48] *J.H. Sohn, M.J. Han, M.Y. Lee, S.K. Kang* and *J.S. Yang:* Simultaneous determination of N-hydroxymethyl-N-methylformamide, N-methylformamide and N-acetyl-S-(N-methylcarbamoyl)cysteine in urine samples from workers exposed to N,N-dimethylformamide by liquid chromatography-tandem mass spectrometry. J. Pharm. Biomed. Anal. 37, 165–170 (2005).

[49] *D. Teshima, N. Kitagawa, K. Otsubo, K. Makino, Y. Itoh* and *R. Oishi:* Simple determination of mycophenolic acid in human serum by column-switching high-performance liquid chromatography. J. Chromatogr. B Analyt. Technol. Biomed. Life Sci. 780, 21–26 (2002).

[50] *J. Ayrton, G.J. Dear, W.J. Leavens, D.N. Mallett* and *R.S. Plumb:* The use of turbulent flow chromatography/mass spectrometry for the rapid, direct analysis of a novel pharmaceutical compound in plasma. Rapid Commun. Mass Spectrom. 11, 1953–1958 (1997).

[51] *N. Brignol, R. Bakhtiar, L. Dou, T. Majumdar* and *F.L. Tse:* Quantitative analysis of terbinafine (Lamisil) in human and minipig plasma by liquid chromatography tandem mass spectrometry. Rapid Commun. Mass Spectrom. 14, 141–149 (2000).

[52] *U. Ceglarek, J. Lembcke, G.M. Fiedler, M. Werner, H. Witzigmann, J.P. Hauss* and *J. Thiery:* Rapid simultaneous quantification of immunosuppressants in transplant patients by turbulent flow chromatography combined with tandem mass spectrometry. Clin. Chim. Acta 346, 181–190 (2004).

[53] *J.L. Herman:* Generic method for on-line extraction of drug substances in the presence of biological matrices using turbulent flow chromatography. Rapid Commun. Mass Spectrom. 16, 421–426 (2002).

[54] *G. Hopfgartner, C. Husser* and *M. Zell:* High-throughput quantification of drugs and their metabolites in biosamples by LC-MS/MS and CE-MS/MS: possibilities and limitations. Ther. Drug Monit. 24, 134–143 (2002).

[55] *J. Smalley, P. Kadiyala, B. Xin, P. Balimane* and *T. Olah:* Development of an on-line extraction turbulent flow chromatography tandem mass spectrometry method for cassette analysis of Caco-2 cell based bi-directional assay samples. J. Chromatogr. B Analyt. Technol. Biomed. Life Sci. 830, 270–277 (2006).

[56] *M. Takino, S. Daishima, K. Yamaguchi* and *T. Nakahara:* Quantitative liquid chromatography-mass spectrometry determination of catechins in human plasma by automated on-line extraction using turbulent flow chromatography. Analyst 128, 46–50 (2003).

[57] *J.T. Wu, H. Zeng, M. Qian, B.L. Brogdon* and *S.E. Unger:* Direct plasma sample injection in multiple-component LC-MS-MS assays for high-throughput pharmacokinetic screening. Anal Chem 72, 61–67 (2000).

[58] *X.S. Xu, K.X. Yan, H. Song* and *M.W. Lo:* Quantitative determination of a novel dual PPAR alpha/gamma agonist using on-line turbulent flow extraction with liquid chromatography-tandem mass spectrometry. J. Chromatogr. B Analyt. Technol. Biomed. Life Sci. 814, 29–36 (2005).

[59] *S. Zhou, H. Zhou, M. Larson, D.L. Miller, D. Mao, X. Jiang* and *W. Naidong:* High-throughput biological sample analysis using on-line turbulent flow extraction combined with monolithic column liquid chromatography/tandem mass spectrometry. Rapid Commun. Mass Spectrom. 19, 2144–2150 (2005).

[60] *J.T. Wu, H. Zeng, M. Qian, B.L. Brogdon* and *S.E. Unger:* Direct plasma sample injection in multiple-component LC-MS-MS assays for high-throughput pharmacokinetic screening. Anal. Chem. 72, 61–67 (2000).

[61] *R.T. Cass, J.S. Villa, D.E. Karr* and *D.E. Schmidt, Jr.:* Rapid bioanalysis of vancomycin in serum and urine by high-performance liquid chromatography tandem mass spectrometry using on-line sample extraction and parallel analytical columns. Rapid Commun. Mass Spectrom. 15, 406–412 (2001).

[62] *M. Jemal, M. Huang, X. Jiang, Y. Mao* and *M.L. Powell:* Direct injection versus liquid-liquid extraction for plasma sample analysis by high performance liquid chromatography with tandem mass spectrometry. Rapid Commun. Mass Spectrom. 13, 2125–2132 (1999).

[63] R. Kahlich, C.H. Gleiter, S. Laufer and B. Kammerer: Quantitative determination of piritramide in human plasma and urine by off- and on-line solid-phase extraction liquid chromatography coupled to tandem mass spectrometry. Rapid Commun. Mass Spectrom. 20, 275–283 (2007).

[64] C.R. Mallet, Z. Lu, J. Mazzeo and U. Neue: Analysis of a basic drug by on-line solid-phase extraction liquid chromatography/tandem mass spectrometry using a mixed mode sorbent. Rapid Commun. Mass Spectrom. 16, 805–813 (2002).

[65] Y.J. Xue, K.C. Turner, J.B. Meeker, J. Pursley, M. Arnold and S. Unger: Quantitative determination of pioglitazone in human serum by direct-injection high-performance liquid chromatography mass spectrometry and its application to a bioequivalence study. J. Chromatogr. B Analyt. Technol. Biomed. Life Sci. 795, 215–226 (2003).

[66] S. Yang, L. Liu, S.L. Gerson and Y. Xu: Measurement of anti-cancer agent methoxyamine in plasma by tandem mass spectrometry with on-line sample extraction. J. Chromatogr. B Analyt. Technol. Biomed. Life Sci. 795, 295–307 (2003).

[67] R. Plumb, G. Dear, D. Mallett and J. Ayrton: Direct analysis of pharmaceutical compounds in human plasma with chromatographic resolution using an alkyl-bonded silica rod column. Rapid Commun. Mass Spectrom. 15, 986–993 (2001).

[68] S. Souverain, S. Rudaz and J.-L. Veuthey: Use of monolithic supports for fast analysis of drugs and metabolites in plasma by direct injection. Chromatographia 57, 569–575 (2003).

[69] K. Kato, M.J. Silva, L.L. Needham and A.M. Calafat: Determination of 16 phthalate metabolites in urine using automated sample preparation and on-line preconcentration/high-performance liquid chromatography/tandem mass spectrometry. Anal. Chem. 77, 2985–2991 (2005).

[70] Y. Alnouti, K. Srinivasan, D. Waddell, H. Bi, O. Kavetskaia and A.I. Gusev: Development and application of a new on-line SPE system combined with LC-MS/MS detection for high throughput direct analysis of pharmaceutical compounds in plasma. J. Chromatogr. A 1080, 99–106 (2005).

[71] W. Weinmann, P. Schaefer, A. Thierauf, A. Schreiber and F.M. Wurst: Confirmatory analysis of ethylglucuronide in urine by liquid-chromatography/electrospray ionization/tandem mass spectrometry according to forensic guidelines. J. Am. Soc. Mass Spectrom. 15, 188–193 (2004).

[72] E.T. Gangl, R.J. Turesky and P. Vouros: Detection of in vivo formed DNA adducts at the part-per-billion level by capillary liquid chromatography/microelectrospray mass spectrometry. Anal. Chem. 73, 2397–2404 (2001).

[73] W. Schütz: Kapillarelektrophorese. Schattauer, Stuttgart; New York (1999).

[74] J.T. Wu, M.G. Qian, M.X. Li, K. Zheng, P. Huang and D.M. Lubman: On-line analysis by capillary separations interfaced to an ion trap storage/reflectron time-of-flight mass spectrometer. J. Chromatogr. A 794, 377–389 (1998).

[75] J. Cai and J. Henion: Capillary electrophoresis-mass spectrometry. J. Chromatogr. A 703, 667–692 (1995).

[76] J. Ding and P. Vouros: Advances in CE/MS. Anal. Chem. 71, 378A–385A (1999).

[77] W.M.A. Niessen, U.R. Tjaden and J. van der Greef: Capillary electrophoresis-mass spectrometry. J. Chromatogr. A 636, 3–19 (1993).

[78] A.J. Tomlinson, N.A. Guzman and S. Naylor: Enhancement of concentration limits of detection in CE and CE-MS: a review of on-line sample extraction, cleanup, analyte preconcentration, and microreactor technology. J. Capillary Electrophor. 2, 247–266 (1995).

[79] W.H. McFadden, H.L. Schwartz and S. Evans: Direct analysis of liquid chromatographic effluents. J. Chromatogr. A 122, 389–396 (1976).

[80] R.P.W. Scott, C.G. Scott, M. Munroe and J. Hess, J.: Interface for on-line liquid chromatography-mass spectroscopy analysis. J. Chromatogr. A 99, 395–405 (1974).

[81] N.J. Alcock, C. Eckers, D.E. Games, M.P.L. Games, M.S. Lant, M.A. McDowall, M. Rossiter, R.W. Smith, S.A. Westwood and H.-Y. Wong: High-performance liquid chromatography-mass spectrometry with transport interfaces. J. Chromatogr. A 251, 165–174 (1982).

[82] D.E. Games, M.A. McDowall, K. Levsen, K.H. Schafer, P. Dobberstein and J.L. Gower: A comparison of moving belt interfaces for LC/MS. Biomed. Mass Spectrom. 11, 87–95 (1984).

[83] R.F. Browner, P.C. Winkler, D.D. Perkins and L.E. Abbey: Aerosols as microsample introduction media for mass spectrometry. Microchem. J. 34, 15–24 (1986).

[84] R.C. Willoughby and R.F. Browner: Monodisperse aerosol generation interface for coupling liquid chromatography with mass spectroscopy. Anal. Chem. 56, 2626–2631 (1984).
[85] T.D. Behymer, T.A. Bellar and W.L. Budde: Liquid chromatography/particle beam mass spectrometry of polar compounds of environmental interest. Anal. Chem. 62, 1686–1690 (1990).
[86] T.A. Bellar, T.D. Behymer and W.L. Budde: Investigation of enhanced ion abundances from a carrier process in high performance liquid chromatography particle beam mass spectrometry. J. Am. Soc. Mass Spectrom. 1, 92–98 (1990).
[87] P. Arpino, M.A. Baldwin and F.W. McLafferty: Liquid chromatography-mass spectrometry. II. Continuous monitoring. Biomed. Mass Spectrom. 1, 80–82 (1974).
[88] M.A. Baldwin and F.W. McLafferty: Liquid chromatography-mass spectrometry interface – I: the direct introduction of liquid solutions into a chemical ionization mass spectrometer. Org. Mass Spectrom. 7, 1111–1112 (1973).
[89] A. Melera: Design, operation and applications of a novel LCMS CI interface. Adv. Mass Spectrom. 8B, 1597–1615 (1980).
[90] W.M.A. Niessen and J. van der Greef: Liquid chromatography/mass spectrometry. Chromatographic Science Series No. 58, Marcel Dekker, New York, USA (1992).
[91] F.R. Sugnaux, D.S. Skrabalak and J.D. Henion: Direct liquid introduction micro-liquid chromatography-mass spectrometry coupling: Optimization of droplet desolvation and instrumental parameters for high sensitivity. J. Chromatogr. A 264, 357–376 (1983).
[92] R.M. Caprioli, T. Fan and J.S. Cottrell: Continuous-flow sample probe for fast atom bombardment mass spectrometry. Anal. Chem. 58, 2949–2954 (1986).
[93] Y. Ito, T. Takeuchi, D. Ishii and M. Goto: Direct coupling of micro high-performance liquid chromatography with fast atom bombardment mass spectrometry. J. Chromatogr. A 346, 161–166 (1985).
[94] R.M.H. Caprioli: Continuous-flow fast atom bombardment mass spectrometry. Wiley, Chichester, England (1990).
[95] J.G. Stroh and K.L. Rinehart: Liquid chromatography/fast atom bombardment mass spectrometry. Top. Mass Spectrom. 1, 287–311 (1994).
[96] C.R. Blakley, J.J. Carmody and M.L. Vestal: A new soft ionization technique for mass spectrometry of complex molecules. J. Am. Chem. Soc. 102, 5931–5933 (1980).
[97] C.R. Blakley, M.J. McAdams and M.L. Vestal: Crossed-beam liquid chromatograph-mass spectrometer combination. J. Chromatogr. A 158, 261–276 (1978).
[98] M.L. Vestal: Ionization techniques for nonvolatile molecules. Mass Spectrom. Rev. 2, 447–480 (1983).
[99] M.L. Vestal and G.J. Fergusson: Thermospray liquid chromatograph/mass spectrometer interface with direct electrical heating of the capillary. Anal. Chem. 57, 2373–2378 (1985).
[100] P. Arpino: Combined liquid chromatography mass spectrometry. Part II. Techniques and mechanisms of thermospray. Mass Spectrom. Rev. 9, 631–669 (1990).
[101] P. Arpino: Combined liquid chromatography mass spectrometry. Part III. Applications of thermospray. Mass Spectrom. Rev. 11, 3–40 (1992).
[102] A.L. Yergey, C.G. Edmonds, A.S. Lewis and M.L. Vestal: Liquid chromatography/mass spectrometry, New York, USA (1990).
[103] S.J. Bos, S.M. van Leeuwen and U. Karst: From fundamentals to applications: recent developments in atmospheric pressure photoionization mass spectrometry. Anal. Bioanal. Chem. 384, 85–99 (2006).
[104] J.B. Fenn, M. Mann, C.K. Meng, S.F. Wong and C.M. Whitehouse: Electrospray ionization for mass spectrometry of large biomolecules. Science 246, 64–71 (1989).
[105] C.M. Whitehouse, R.N. Dreyer, M. Yamashita and J.B. Fenn: Electrospray interface for liquid chromatographs and mass spectrometers. Anal. Chem. 57, 675–679 (1985).
[106] M. Yamashita and J.B. Fenn: Electrospray ion source. Another variation on the free-jet theme. J. Phys. Chem. 88, 4451–4459 (1984).
[107] M.L. Aleksandrov, G.I. Barama, L.M. Gall, N.V. Krasnov, Y.S. Kusner, O.A. Irgorodskaya, V.I. Nikolaiev and V.A. Shkurov: Use of a novel mass-spectrometric method for the sequencing of peptides. Bioorg. Khim. 11, 700–705 (1985).

[108] P. Kebarle: A brief overview of the present status of the mechanisms involved in electrospray mass spectrometry. J. Mass Spectrom. 35, 804–817 (2000).
[109] M.G. Ikonomou, A.T. Blades and P. Kebarle: Electrospray mass spectrometry of methanol and water solutions suppression of electric discharge with SF6 gas. J. Am. Soc. Mass Spectrom. 2, 497–505 (1991).
[110] D. Duft, T. Achtzehn, R. Muller, B.A. Huber and T. Leisner: Coulomb fission: Rayleigh jets from levitated microdroplets. Nature 421, 128 (2003).
[111] A. Gomez and K. Tang: Charge and fission of droplets in electrostatic sprays. Phys. Fluids 6, 404–414 (1994).
[112] H. Nehring, S. Thiebes, L. Bütfering and F.W. Röllgen: Cluster ion formation in thermospray mass spectrometry of ammonium salts. Int. J. Mass Spectrom. Ion Processes 128, 123–132 (1993).
[113] G. Schmelzeisen-Redeker, L. Bütfering and F.W. Röllgen: Desolvation of ions and molecules in thermospray mass spectrometry. Int. J. Mass Spectrom. Ion Processes 90, 139–150 (1989).
[114] J.V. Iribarne and B.A. Thomson: On the evaporation of small ions from charged droplets. J. Chem. Phys. 64, 2287–2294 (1976).
[115] B.A. Thomson and J.V. Iribarne: Field induced ion evaporation from liquid surfaces at atmospheric pressure. J. Chem. Phys. 71, 4451–4463 (1979).
[116] R.D. Smith, J.A. Loo, C.G. Edmonds, C.J. Barinaga and H.R. Udseth: New developments in biochemical mass spectrometry: electrospray ionization. Anal. Chem. 62, 882–899 (1990).
[117] W.M. Niessen: Advances in instrumentation in liquid chromatography-mass spectrometry and related liquid-introduction techniques. J. Chromatogr. A 794, 407–435 (1998).
[118] A.P. Bruins, T.R. Covey and J.D. Henion: Ion spray interface for combined liquid chromatography/atmospheric pressure ionization mass spectrometry. Anal. Chem. 59, 2642–2646 (1987).
[119] J.F. Banks, Jr., S. Shen, C.M. Whitehouse and J.B. Fenn: Ultrasonically assisted electrospray ionization for LC/MS determination of nucleosides from a transfer RNA digest. Anal. Chem. 66, 406–414 (1994).
[120] T.R. Covey, E.D. Lee and J.D. Henion: High-speed liquid chromatography/tandem mass spectrometry for the determination of drugs in biological samples. Anal. Chem. 58, 2453–2460 (1986).
[121] E.C. Horning, D.I. Carroll, I. Dzidic, K.D. Haegele, M.G. Horning and R.N. Stillwell: Atmospheric pressure ionization (API) mass spectrometry. Solvent-mediated ionization of samples introduced in solution and in a liquid chromatograph effluent stream. J. Chromatogr. Sci. 12, 725–729 (1974).
[122] E.C. Horning, M.G. Horning, D.I. Carroll, I. Dzidic and R.N. Stillwell: New picogram detection system based on a mass spectrometer with an external ionization source at atmospheric pressure. Anal. Chem. 45, 936–943 (1973).
[123] M.L. Huertas and J. Fontan: Evolution times of the tropospheric positive ions. Atmospheric Environ. 9, 1018 (1975).
[124] D.I. Carroll, I. Dzidic, E.C. Horning and R.N. Stillwell: Atmospheric pressure ionization mass spectrometry. Appl. Spectros. Rev. 17, 337–406 (1981).
[125] G. Nicol, J. Sunner and P. Kebarle: Kinetics and thermodynamics of protonation reactions: $H_3O + (H_2O)_h$ (hydronium hydrate) + B = $BH^+ (H_2O)_b$ + (h – b + 1) H_2O where B is a nitrogen, oxygen, or carbon base. Int. J. Mass Spectrom. Ion Processes 84, 135–155 (1988).
[126] J. Sunner, M.G. Ikonomou and P. Kebarle: Sensitivity enhancements obtained at high temperatures in atmospheric pressure ionization mass spectrometry. Anal. Chem. 60, 1308–1313 (1988).
[127] J. Sunner, G. Nicol and P. Kebarle: Factors determining relative sensitivity of analytes in positive mode atmospheric pressure ionization mass spectrometry. Anal. Chem. 60, 1300–1307 (1988).
[128] D.B. Robb, T.R. Covey and A.P. Bruins: Atmospheric pressure photoionization: an ionization method for liquid chromatography-mass spectrometry. Anal. Chem. 72, 3653–3659 (2000).
[129] J.A. Syage: Mechanism of [M + H]+ formation in photoionization mass spectrometry. J. Am. Soc. Mass Spectrom. 15, 1521–1533 (2004).

[130] *P. Traldi:* Atmospheric pressure photoionization mass spectrometry. Mass Spectrom. Rev. 22, 318–331 (2003).
[131] *E. Marotta, R. Seraglia, F. Fabris* and *P. Traldi:* Atmospheric pressure photoionization mechanisms: 1. The case of acetonitrile. Int. J. Mass Spectrom. Ion Processes 228, 841–849 (2003).
[132] *E. Marotta* and *P. Traldi:* On the photo-initiated isomerization of acetonitrile. Rapid Commun. Mass Spectrom. 17, 2846–2848 (2003).
[133] *E. Straube, W. Dekant* and *W. Völkel:* Enhanced sensitivity for the determination of ambiphilic polyaromatic amines by LC-MS/MS after acetylation. J. Chromatogr. A 1067, 181–190 (2005).
[134] *E. A. Straube, W. Dekant* and *W. Völkel:* Comparison of electrospray ionization, atmospheric pressure chemical ionization, and atmospheric pressure photoionization for the analysis of dinitropyrene and aminonitropyrene LC-MS/MS. J. Am. Soc. Mass Spectrom. 15, 1853–1862 (2004).
[135] *C. Dass:* Mass spectrometry: clinical and biomedical applications. Plenum Press, New York, USA (1994).
[136] *P. E. Miller* and *M. B. Denton:* The quadrupole mass filter: basic operating concepts. J. Chem. Educ. 7, 617–622 (1986).
[137] *R. E. March* and *R. J. Hughes:* Quadrupole storage MS. John Wiley and Sons, New York (1989).
[138] *W. E. Stephens:* A pulsed mass spectrometer with time dispersion. Phys. Rev. 69, 691 (1946).
[139] *W. C. Wiley* and *I. H. McLaren:* Time-of-flight mass spectrometer with improved resolution. Rev. Sci. Instr. 26, 1150–1157 (1955).
[140] *R. J. Cotter:* Time-of-flight mass spectrometry. ACS, Washington D.C. (1994).
[141] *I. V. Chernushevich, W. Ens* and *K. G. Standing:* Orthogonal-injection TOFMS for analyzing biomolecules. Anal. Chem. 71, 452A–461A (1999).
[142] *R. J. Cotter:* The new time-of-flight mass spectrometry. Anal. Chem. 71, 445A–451A (1999).
[143] *C. M. Henry:* Electrospray in flight. Orthogonal acceleration brings the advantages of time of flight to electrospray. Anal. Chem. 71, 197A–201A (1999).
[144] *B. A. Mamyrin, V. I. Karataev, D. V. Shmikk* and *V. A. Zagulin:* The mass-reflectron, a new nonmagnetic time-of-flight mass spectrometer with high resolution. Sov. Phys. JETP 37, 45 (1973).
[145] *E. O. Lawrence* and *M. S. Livingston:* The production of high speed light ions without the use of high voltages. Phys. Rev. 40, 19–35 (1932).
[146] *M. B. Comisarow* and *A. G. Marshall:* Fourier transform ion cyclotron resonance spectroscopy. Chem. Phys. Lett. 25, 282–283 (1974).
[147] *A. G. Marshall:* Milestones in Fourier transform ion cyclotron resonance mass spectrometry technique development. Int. J. Mass Spectrom. Ion Processes 200, 331–356 (2000).
[148] *A. Makarov, E. Denisov, A. Kholomeev, W. Balschun, O. Lange, K. Strupat* and *S. Horning:* Performance evaluation of a hybrid linear ion trap/orbitrap mass spectrometer. Anal. Chem. 78, 2113–2120 (2006).
[149] *Q. Hu, R. J. Noll, H. Li, A. Makarov, M. Hardman* and *R. Graham Cooks:* The Orbitrap: a new mass spectrometer. J. Mass Spectrom. 40, 430–443 (2005).
[150] *B. Erickson:* Linear ion trap/Orbitrap mass spectrometer. Anal. Chem. 78, 2089 (2006).
[151] *A. Makarov, E. Denisov, O. Lange* and *S. Horning:* Dynamic range of mass accuracy in LTQ Orbitrap hybrid mass spectrometer. J. Am. Soc. Mass Spectrom. 17, 977–982 (2006).
[152] *G. Hopfgartner, C. Husser* and *M. Zell:* Rapid screening and characterization of drug metabolites using a new quadrupole-linear ion trap mass spectrometer. J. Mass Spectrom. 38, 138–150 (2003).
[153] *G. Hopfgartner, E. Varesio, V. Tschappat, C. Grivet, E. Bourgogne* and *L. A. Leuthold:* Triple quadrupole linear ion trap mass spectrometer for the analysis of small molecules and macromolecules. J. Mass Spectrom. 39, 845–855 (2004).
[154] *L. A. Leuthold, C. Grivet, M. Allen, M. Baumert* and *G. Hopfgartner:* Simultaneous selected reaction monitoring, MS/MS and MS3 quantitation for the analysis of pharmaceutical compounds in human plasma using chip-based infusion. Rapid Commun. Mass Spectrom. 18, 1995–2000 (2004).
[155] *S. Marchese, R. Curini, A. Gentili, D. Perret* and *L. M. Rocca:* Simultaneous determination of the urinary metabolites of benzene, toluene, xylene and styrene using high-performance liquid

chromatography/hybrid quadrupole time-of-flight mass spectrometry. Rapid Commun. Mass Spectrom. 18, 265–272 (2004).

[156] *C. Soler, B. Hamilton, A. Furey, K. J. James, J. Manes* and *Y. Pico:* Comparison of four mass analyzers for determining carbosulfan and its metabolites in citrus by liquid chromatography/ mass spectrometry. Rapid Commun. Mass Spectrom. 20, 2151–2164 (2006).

[157] *R. F. Staack, E. Varesio* and *G. Hopfgartner:* The combination of liquid chromatography/tandem mass spectrometry and chip-based infusion for improved screening and characterization of drug metabolites. Rapid Commun. Mass Spectrom. 19, 618–626 (2005).

[158] *S. M. van Leeuwen, L. Hendriksen* and *U. Karst:* Determination of aldehydes and ketones using derivatization with 2,4-dinitrophenylhydrazine and liquid chromatography-atmospheric pressure photoionization-mass spectrometry. J. Chromatogr. A 1058, 107–112 (2004).

[159] *G. Zurek, A. Buldt* and *U. Karst:* Determination of acetaldehyde in tobacco smoke using N-methyl-4-hydrazino-7-nitrobenzofurazan and liquid chromatography/mass spectrometry. Fresenius J. Anal. Chem. 366, 396–399 (2000).

[160] *X. Ye, Z. Kuklenyik, L. L. Needham* and *A. M. Calafat:* Measuring environmental phenols and chlorinated organic chemicals in breast milk using automated on-line column-switching-high performance liquid chromatography-isotope dilution tandem mass spectrometry. J. Chromatogr. B Analyt. Technol. Biomed. Life Sci. 831, 110–115 (2006).

[161] *X. Ye, Z. Kuklenyik, L. L. Needham* and *A. M. Calafat:* Automated on-line column-switching HPLC-MS/MS method with peak focusing for the determination of nine environmental phenols in urine. Anal. Chem. 77, 5407–5413 (2005).

[162] *M. I. Boettcher* and *J. Angerer:* Determination of the major mercapturic acids of acrylamide and glycidamide in human urine by LC-ESI-MS/MS. J. Chromatogr. B Analyt. Technol. Biomed. Life Sci. 824, 283–294 (2005).

[163] *C. M. Li, C. W. Hu* and *K. Y. Wu:* Quantification of urinary N-acetyl-S- (propionamide)cysteine using an on-line clean-up system coupled with liquid chromatography/tandem mass spectrometry. J. Mass Spectrom. 40, 511–515 (2005).

[164] *M. Urban, D. Kavvadias, K. Riedel, G. Scherer* and *A. R. Tricker:* Urinary mercapturic acids and a hemoglobin adduct for the dosimetry of acrylamide exposure in smokers and nonsmokers. Inhal. Toxicol. 18, 831–839 (2006).

[165] *S. Marchese, R. Curini, A. Gentili, D. Perret* and *L. M. Rocca:* Simultaneous determination of the urinary metabolites of benzene, toluene, xylene and styrene using high-performance liquid chromatography/hybrid quadrupole time-of-flight mass spectrometry. Rapid Commun. Mass Spectrom. 18, 265–272 (2004).

[166] *C. E. Garner, S. C. Sumner, J. G. Davis, J. P. Burgess, Y. Yueh, J. Demeter, Q. Zhan, J. Valentine, A. R. Jeffcoat, L. T. Burka* and *J. M. Mathews:* Metabolism and disposition of 1-bromopropane in rats and mice following inhalation or intravenous administration. Toxicol. Appl. Pharmacol. 215, 23–36 (2006).

[167] *M. Urban, G. Gilch, G. Schepers, E. van Miert* and *G. Scherer:* Determination of the major mercapturic acids of 1,3-butadiene in human and rat urine using liquid chromatography with tandem mass spectrometry. J. Chromatogr. B Analyt. Technol. Biomed. Life Sci. 796, 131–140 (2003).

[168] *P. Manini, R. Andreoli, E. Bergamaschi, G. De Palma, A. Mutti* and *W. M. Niessen:* A new method for the analysis of styrene mercapturic acids by liquid chromatography/electrospray tandem mass spectrometry. Rapid Commun. Mass Spectrom. 14, 2055–2060 (2000).

[169] *D. B. Barr* and *D. L. Ashley:* A rapid, sensitive method for the quantitation of N-acetyl-S-(2-hydroxyethyl)-L-cysteine in human urine using isotope-dilution HPLC-MS-MS. J. Anal. Toxicol. 22, 96–104 (1998).

[170] *A. M. Calafat, D. B. Barr, J. L. Pirkle* and *D. L. Ashley:* Reference range concentrations of N-acetyl-S-(2-hydroxyethyl)-L-cysteine, a common metabolite of several volatile organic compounds, in the urine of adults in the United States. J. Expo. Anal. Environ. Epidemiol. 9, 336–342 (1999).

[171] L. M. Gonzalez-Reche, T. Schettgen and J. Angerer: New approaches to the metabolism of xylenes: verification of the formation of phenylmercapturic acid metabolites of xylenes. Arch. Toxicol. 77, 80–85 (2003).

[172] R. J. Turesky, F. P. Guengerich, A. Guillouzo and S. Langouet: Metabolism of heterocyclic aromatic amines by human hepatocytes and cytochrome P4501A2. Mutat. Res. 506–507, 187–195 (2002).

[173] K. Inoue, M. Kawaguchi, Y. Funakoshi and H. Nakazawa: Size-exclusion flow extraction of bisphenol A in human urine for liquid chromatography-mass spectrometry. J. Chromatogr. B Analyt. Technol. Biomed. Life Sci. 798, 17–23 (2003).

[174] X. Chen, F. Qiu, D. Zhong, X. Duan and C. Liu: Validated liquid chromatography-tandem mass spectrometric method for the quantitative determination of daidzein and its main metabolite daidzein glucuronide in rat plasma. Pharmazie 60, 334–338 (2005).

[175] W. Bicker, M. Lammerhofer, T. Keller, R. Schuhmacher, R. Krska and W. Lindner: Validated method for the determination of the ethanol consumption markers ethyl glucuronide, ethyl phosphate, and ethyl sulfate in human urine by reversed-phase/weak anion exchange liquid chromatography-tandem mass spectrometry. Anal. Chem. 78, 5884–5892 (2006).

[176] S. Dresen, W. Weinmann and F. M. Wurst: Forensic confirmatory analysis of ethyl sulfate – a new marker for alcohol consumption – by liquid-chromatography/electrospray ionization/tandem mass spectrometry. J. Am. Soc. Mass Spectrom. 15, 1644–1648 (2004).

[177] F. M. Wurst, S. Dresen, J. P. Allen, G. Wiesbeck, M. Graf and W. Weinmann: Ethyl sulphate: a direct ethanol metabolite reflecting recent alcohol consumption. Addiction 101, 204–211 (2006).

[178] P. Manini, R. Andreoli, A. Mutti, E. Bergamaschi and I. Franchini: Determination of free and glucuronated hexane metabolites without prior hydrolysis by liquid- and gas-chromatography coupled with mass spectrometry. Toxicol. Lett. 108, 225–231 (1999).

[179] P. Manini, R. Andreoli, A. Mutti, E. Bergamaschi and W. M. Niessen: Determination of n-hexane metabolites by liquid chromatography/mass spectrometry. 2. Glucuronide-conjugated metabolites in untreated urine samples by electrospray ionization. Rapid Commun. Mass Spectrom. 12, 1615–1624 (1998).

[180] D. L. Heavner, J. D. Richardson, W. T. Morgan and M. W. Ogden: Validation and application of a method for the determination of nicotine and five major metabolites in smokers' urine by solid-phase extraction and liquid chromatography-tandem mass spectrometry. Biomed. Chromatogr. 19, 312–328 (2005).

[181] Y. Li, A. C. Li, H. Shi, S. Zhou, W. Z. Shou, X. Jiang, W. Naidong and J. H. Lauterbach: The use of chemical derivatization to enhance liquid chromatography/tandem mass spectrometric determination of 1-hydroxypyrene, a biomarker for polycyclic aromatic hydrocarbons in human urine. Rapid Commun. Mass Spectrom. 19, 3331–3338 (2005).

[182] A. O. Olsson, S. E. Baker, J. V. Nguyen, L. C. Romanoff, S. O. Udunka, R. D. Walker, K. L. Flemmen and D. B. Barr: A liquid chromatography – tandem mass spectrometry multiresidue method for quantification of specific metabolites of organophosphorus pesticides, synthetic pyrethroids, selected herbicides, and DEET in human urine. Anal. Chem. 76, 2453–2461 (2004).

[183] B. C. Blount, K. E. Milgram, M. J. Silva, N. A. Malek, J. A. Reidy, L. L. Needham and J. W. Brock: Quantitative detection of eight phthalate metabolites in human urine using HPLC-APCI-MS/MS. Anal. Chem. 72, 4127–4134 (2000).

[184] B. C. Blount, M. J. Silva, S. P. Caudill, L. L. Needham, J. L. Pirkle, E. J. Sampson, G. W. Lucier, R. J. Jackson and J. W. Brock: Levels of seven urinary phthalate metabolites in a human reference population. Environ. Health Perspect. 108, 979–982 (2000).

[185] S. M. Duty, M. J. Silva, D. B. Barr, J. W. Brock, L. Ryan, Z. Chen, R. F. Herrick, D. C. Christiani and R. Hauser: Phthalate exposure and human semen parameters. Epidemiology 14, 269–277 (2003).

[186] K. Kato, M. J. Silva, J. W. Brock, J. A. Reidy, N. A. Malek, C. C. Hodge, H. Nakazawa, L. L. Needham and D. B. Barr: Quantitative detection of nine phthalate metabolites in human serum using reversed-phase high-performance liquid chromatography-electrospray ionization-tandem mass spectrometry. J. Anal. Toxicol. 27, 284–289 (2003).

[187] *M.J. Silva, N.A. Malek, C.C. Hodge, J.A. Reidy, K. Kato, D.B. Barr, L.L. Needham* and *J.W. Brock:* Improved quantitative detection of 11 urinary phthalate metabolites in humans using liquid chromatography-atmospheric pressure chemical ionization tandem mass spectrometry. J. Chromatogr. B Analyt. Technol. Biomed. Life Sci. 789, 393–404 (2003).

[188] *U.M. Schauer, W. Völkel* and *W. Dekant:* Toxicokinetics of tetrabromobisphenol a in humans and rats after oral administration. Toxicol. Sci. 91, 49–58 (2006).

[189] *D.B. Barr, W.E. Turner, E. DiPietro, P.C. McClure, S.E. Baker, J.R. Barr, K. Gehle, R.E. Grissom, Jr., R. Bravo, W.J. Driskell, D.G. Patterson, Jr., R.H. Hill, Jr., L.L. Needham, J.L. Pirkle* and *E.J. Sampson:* Measurement of p-nitrophenol in the urine of residents whose homes were contaminated with methyl parathion. Environ. Health Perspect. 110 Suppl. 6, 1085–1091 (2002).

[190] *A.O. Olsson, J.V. Nguyen, M.A. Sadowski* and *D.B. Barr:* A liquid chromatography/electrospray ionization-tandem mass spectrometry method for quantification of specific organophosphorus pesticide biomarkers in human urine. Anal. Bioanal. Chem. 376, 808–815 (2003).

[191] *A.D. Delinsky, D.C. Delinsky, S. Muralidhara, J.W. Fisher, J.V. Bruckner* and *M.G. Bartlett:* Analysis of dichloroacetic acid in rat blood and tissues by hydrophilic interaction liquid chromatography with tandem mass spectrometry. Rapid Commun. Mass Spectrom. 19, 1075–1083 (2005).

[192] *X. Ye, A.M. Bishop, J.A. Reidy, L.L. Needham* and *A.M. Calafat:* Parabens as urinary biomarkers of exposure in humans. Environ. Health Perspect. 114, 1843–1846 (2006).

[193] *X. Ye, Z. Kuklenyik, A.M. Bishop, L.L. Needham* and *A.M. Calafat:* Quantification of the urinary concentrations of parabens in humans by on-line solid phase extraction-high performance liquid chromatography-isotope dilution tandem mass spectrometry. J. Chromatogr. B Analyt. Technol. Biomed. Life Sci. 844, 53–59 (2006).

[194] *H. Fromme, O. Midasch, D. Twardella, J. Angerer, S. Boehmer* and *B. Liebl:* Occurrence of perfluorinated substances in an adult German population in southern Bavaria. Int. Arch. Occup. Environ. Health 80(4), 313–319 (2006).

[195] *Z. Kuklenyik, J.A. Reich, J.S. Tully, L.L. Needham* and *A.M. Calafat:* Automated solid-phase extraction and measurement of perfluorinated organic acids and amides in human serum and milk. Environ. Sci. Technol. 38, 3698–3704 (2004).

[196] *O. Midasch, T. Schettgen* and *J. Angerer:* Pilot study on the perfluorooctanesulfonate and perfluorooctanoate exposure of the German general population. Int. J. Hyg. Environ. Health 209, 489–496 (2006).

[197] *R. Preuss, H.M. Koch* and *J. Angerer:* Biological monitoring of the five major metabolites of di-(2-ethylhexyl)phthalate (DEHP) in human urine using column-switching liquid chromatography-tandem mass spectrometry. J. Chromatogr. B Analyt. Technol. Biomed. Life Sci. 816, 269–280 (2005).

[198] *H.M. Koch, L.M. Gonzalez-Reche* and *J. Angerer:* On-line clean-up by multidimensional liquid chromatography-electrospray ionization tandem mass spectrometry for high throughput quantification of primary and secondary phthalate metabolites in human urine. J Chromatogr B Analyt Technol Biomed Life Sci 784, 169–182 (2003).

[199] *B.C. Blount, K.E. Milgram, M.J. Silva, N.A. Malek, J.A. Reidy, L.L. Needham* and *J.W. Brock:* Quantitative detection of eight phthalate metabolites in human urine using HPLC-APCI-MS/MS. Anal. Chem. 72, 4127–4134 (2000).

[200] *R.J. Turesky, J.P. Freeman, R.D. Holland, D.M. Nestorick, D.W. Miller, D.L. Ratnasinghe* and *F.F. Kadlubar:* Identification of aminobiphenyl derivatives in commercial hair dyes. Chem. Res. Toxicol. 16, 1162–1173 (2003).

[201] *R.D. Holland, T. Gehring, J. Taylor, B.G. Lake, N.J. Gooderham* and *R.J. Turesky:* Formation of a mutagenic heterocyclic aromatic amine from creatinine in urine of meat eaters and vegetarians. Chem. Res. Toxicol. 18, 579–590 (2005).

[202] *R.D. Holland, J. Taylor, L. Schoenbachler, R.C. Jones, J.P. Freeman, D.W. Miller, B.G. Lake, N.J. Gooderham* and *R.J. Turesky:* Rapid biomonitoring of heterocyclic aromatic amines in human urine by tandem solvent solid phase extraction liquid chromatography electrospray ionization mass spectrometry. Chem. Res. Toxicol. 17, 1121–1136 (2004).

[203] C. W. Hu, M. T. Wu, M. R. Chao, C. H. Pan, C. J. Wang, J. A. Swenberg and K. Y. Wu: Comparison of analyses of urinary 8-hydroxy-2′-deoxyguanosine by isotope-dilution liquid chromatography with electrospray tandem mass spectrometry and by enzyme-linked immunosorbent assay. Rapid Commun. Mass Spectrom. 18, 505–510 (2004).
[204] U. Lutz, W. Völkel, R. W. Lutz and W. K. Lutz: LC-MS/MS analysis of dextromethorphan metabolism in human saliva and urine to determine CYP2D6 phenotype and individual variability in N-demethylation and glucuronidation. J. Chromatogr. B Analyt. Technol. Biomed. Life Sci. 813, 217–225 (2004).
[205] N. Picard, D. Ratanasavanh, A. Premaud, Y. Le Meur and P. Marquet: Identification of the UDP-glucuronosyltransferase isoforms involved in mycophenolic acid phase II metabolism. Drug Metab. Dispos. 33, 139–146 (2005).
[206] W. Völkel, T. Colnot, G. A. Csanady, J. G. Filser and W. Dekant: Metabolism and kinetics of bisphenol a in humans at low doses following oral administration. Chem. Res. Toxicol. 15, 1281–1287 (2002).

Authors: *M. Müller, W. Völkel*
With contributions from: *J. Angerer, H.M. Koch* and *G. Scherer*

Methods

Bisphenol A

Application   Determination in urine

Analytical principle   Capillary gas chromatography/
                            mass spectrometric detection (MS)

Summary

The procedure described here permits the assay of bisphenol A in urine. Exposure to bisphenol A of relevance to both occupational and environmental medicine can be determined.

After addition of the internal standard D_8-bisphenol A, the urine is initially subjected to enzymatic hydrolysis. The analyte is subsequently enriched by means of solid phase extraction, while being simultaneously separated from matrix components. After elution, the analyte is converted to its derivative using pentafluorobenzoyl chloride. After separation by capillary gas chromatography, the derivative is measured using a mass selective detector and electron impact ionisation (EI).

Calibration curves are plotted for quantitative evaluation using calibration standard solutions prepared by spiking pooled human urine with standard substance. The calibration standards are treated in the same manner as the urine samples to be investigated.

Bisphenol A

Within-series imprecision: Standard deviation (rel.) $s_w = 4.1\%$ or 2.9%
                            Prognostic range $u = 10.1\%$ or 7.2%
                            at a spiked concentration of 1 or 10 µg per litre urine
                            and where n = 6 determinations.

Between-day imprecision: Standard deviation (rel.) $s_w = 6.7\%$ or 3.9%
                            Prognostic range $u = 16.5\%$ or 9.6%
                            at a spiked concentration of 1 and 10 µg per litre urine
                            and where n = 6 determinations.

Accuracy: Recovery rate $r = 87\%$ at 1 µg/L and
                                                        104% at 10 µg/L

Quantitation limit: 1 µg bisphenol A per litre urine

The MAK-Collection Part IV: Biomonitoring Methods, Vol. 11.
DFG, Deutsche Forschungsgemeinschaft
Copyright © 2008 WILEY-VCH Verlag GmbH & Co. KGaA, Weinheim
ISBN: 978-3-527-31596-3

Bisphenol A

Bisphenol A (4,4′-isopropylidenediphenol, 2,2-bis-(4-hydroxyphenyl)propane; CAS No. 80-05-7; molecular weight 228.3 g/mol) is mainly used as an intermediate product in the production of epoxy, polycarbonate and styrene-polyester resins. In addition, bisphenol A serves as an antioxidant and a biocide. Exposure to bisphenol A can occur at the workplace in manufacturing and processing plants as well as in areas of interest to environmental medicine (e.g. from polycarbonate bottles or food cans).

It has been found that oral administration of 5 mg bisphenol A to humans (equivalent to 54 to 90 µg/kg body weight) resulted in terminal elimination half-lives of 5.3 hours in plasma (maximum concentration was reached after 80 min) and 5.4 h in urine. After 24 to 34 hours the concentrations had fallen below the detection limit [1]. Bisphenol A is almost exclusively excreted as a glucuronic acid conjugate in man and its elimination is considerably more rapid than in the case of the rat (enterohepatic cycle) [1, 2].

A MAK value (maximum permissible concentration at the workplace) of 5 mg/m^3 (measured as the inhalable fraction) was assigned by the Commission [3, 4]. A comprehensive treatise on the toxicological aspects of bisphenol A can be found in the MAK value documentation [4].

As the concentration of free bisphenol A in blood is toxicologically relevant, its detection in this matrix would be suitable for risk evaluation. However, on account of the rapid metabolisation of bisphenol A, detection in blood is only meaningful directly (1 to 3 hours) after high exposure. Detoxified inactive bisphenol A can be detected in urine up to about 2 days after exposure [2]. Therefore detection of total bisphenol A in urine is preferable in order to assess the total exposure of interest to occupational and environmental medicine. Table 1 presents an overview of the bis-

Table 1. Data on the internal exposure to bisphenol A of people who encounter the substance at the workplace

References	n	Matrix	Free bisphenol A [µg/L]	Total bisphenol A [µg/L]
Sajika et al. (1999) [5]	21	Serum	<0.2–1.6	–
Brock et al. (2001) [2]	5	Urine	<0.12	0.1–0.5
Tsukioka et al. (2003) [6]	6	Urine	–	0.2–3.8
Imai and Morita (2001) [7]	12	Urine	0.008–0.06	0.06–2.3
Matsumoto et al. (2003) [8]	50	Urine	<1.7	<10

Table 2. Data on the internal exposure to bisphenol A of people who do not handle the substance at the workplace

Number n	<DL* [%]	Mean value [µg/L urine]	Median [µg/L urine]	95th percentile [µg/L urine]	Range [µg/L urine]
15	87	0.27	0.21	0.68	<0.1–0.86

* DL = detection limit

phenol A concentrations detected in investigations carried out in the field of occupational medicine.

A background exposure to bisphenol A is detectable in humans due to its ubiquitous distribution. According to Völkel et al. (2002) intake from the environment amounts to 9 µg/kg body weight/day [1]. The results of an investigation carried out as part of the method evaluation on urine samples of persons who do not handle bisphenol A at the workplace are given in Table 2.

Authors: *G. Leng, W. Gries*
Examiner: *M. Blaszkewicz*

Bisphenol A

Application Determination in urine

Analytical principle Capillary gas chromatography/ mass spectrometric detection (MS)

Contents

1 General principles
2 Equipment, chemicals and solutions
2.1 Equipment
2.2 Chemicals
2.3 Solutions
2.4 Calibration standards
2.5 Conditioning the RP 18 columns
3 Specimen collection and sample preparation
3.1 Sample preparation
4 Operational parameters
4.1 Operational parameters for gas chromatography
4.2 Operational parameters for mass spectrometry
5 Analytical determination
6 Calibration
7 Calculation of the analytical result
8 Standardisation and quality control
9 Evaluation of the method
9.1 Precision
9.2 Accuracy
9.3 Detection limits
9.4 Sources of error
10 Discussion of the method
11 References

1 General principles

After addition of the internal standard D_8-bisphenol A, the urine is initially subjected to enzymatic hydrolysis. The analyte is subsequently enriched by means of solid phase extraction, while being simultaneously separated from matrix components. After elution, the analyte is converted to its derivative using pentafluorobenzoyl chloride. After separation by capillary gas chromatography, the derivative is measured using a mass selective detector and electron impact ionisation (EI).

Calibration curves are plotted for quantitative evaluation using calibration standard solutions prepared by spiking pooled human urine with standard substance. The standard solutions are treated in the same manner as the urine samples to be investigated.

2 Equipment, chemicals and solutions

2.1 Equipment

Gas chromatograph with split/splitless injector, mass selective detector, autosampler and data processing system

Capillary gas chromatographic column:
Length: 25 m; inner diameter: 0.2 mm; stationary phase: 100% dimethylpolysiloxane; film thickness: 0.11 μm (e.g. HP 1, Agilent Technologies, No. 19091Z-002)

100 mL Urine collection vessels made of glass (e.g. Schott, No. 215-8401)

100 mL and 500 mL Glass beakers

100 mL Screw-capped jars with lids

Glass Pasteur pipettes

Analytical balance

Laboratory centrifuge

10 mL, 100 mL and 500 mL Volumetric flasks

12 mL Test-tubes with Teflon-coated screw caps (e.g. Schütt No. 3.561.103)

Incubation cupboard (e.g. from Heraeus)

Variably adjustable pipettes, 10 to 100 μL and 100 to 1000 μL (e.g. Transferpette, from Brand)

200 μL Vials (e.g. from Macherey-Nagel)

Crimp caps (e.g. from Macherey-Nagel)

Shaker (e.g. IKA Vibrax VXR)

pH Meter

Magnetic stirrer with stirring rod

RP 18 cartridges (e.g. Phenomenex Strata C18-E, No. 8B-S001-HCH)

Bisphenol A

2.2 Chemicals

iso-Octane SupraSolv®, (e.g. Merck, No. 1.15440.1000)

Methanol SupraSolv® (e.g. Merck, No. 1.06011.2500)

Acetonitrile SupraSolv® (e.g. Merck, No. 1.00017.2500)

Sodium acetate trihydrate, p.a. (e.g. Merck, No. 1.06267.0500)

Hydrochloric acid, fuming, 37%, p.a. (e.g. Merck, No. 1.00317.1000)

Sodium hydrogen carbonate, p.a. (e.g. Merck, No. 1.06329.0500)

Deionised water (e.g. Milli-Q water)

Pentafluorobenzoyl chloride, ultrapure ≥99.0% (e.g. Fluka, No. 76733)

Glucuronidase/arylsulphatase (e.g. Roche, No. 90218220)

Helium 5.0 (e.g. from Linde)

Bisphenol A, pure, ≥95.0% (e.g. Fluka, No. 14939)

Bisphenol A Ring-D_8, 98% (e.g. Cambridge Isotope Laboratories, No. DLM-1540-0.5)

2.3 Solutions

0.1 M Sodium acetate buffer (pH 5):
6.8 g Sodium acetate trihydrate are weighed in a 500 mL glass beaker and dissolved in approx. 300 mL deionised water. Using a pH meter, the pH is adjusted to a value of 5 by adding conc. hydrochloric acid drop by drop while the contents are being stirred constantly. Then the contents are transferred to a 500 mL volumetric flask and the flask is filled to its nominal volume with deionised water.

Elution solution (acetonitrile/methanol, 75:25, v:v):
75 mL acetonitrile and 25 mL methanol are measured using a measuring cylinder and then thoroughly mixed in a 100 mL screw-capped jar.

1 M Sodium hydrogen carbonate solution:
8.4 g sodium hydrogen carbonate are weighed in a 100 mL glass beaker, dissolved in approx. 50 mL deionised water and then transferred to a 100 mL volumetric flask. The flask is subsequently filled to its nominal volume with deionised water.

These solutions can be stored for 6 months at +4 to +6 °C.

2.4 Calibration standards

Stock solution:
10.0 mg bisphenol A are weighed exactly into a 10 mL volumetric flask and dissolved in acetonitrile. The flask is subsequently filled to its nominal volume with acetonitrile. The concentration of this solution is 1 g/L.

Working solution A:
0.1 mL of the stock solution is pipetted into a 100 mL volumetric flask and the flask is filled to its nominal value with acetonitrile. The concentration of this solution is 1 mg/L.

Working solution B:
1 mL of working solution A is pipetted into a 10 mL volumetric flask and the flask is filled to its nominal value with acetonitrile. The concentration of this solution is 0.1 mg/L.

Working solution C:
0.1 mL of working solution A is pipetted into a 10 mL volumetric flask and the flask is filled to its nominal value with acetonitrile. The concentration of this solution is 0.01 mg/L.

Solution of the internal standard (ISTD)

Stock solution of the internal standard:
10 mg D_8-bisphenol A are weighed exactly into a 10 mL volumetric flask and dissolved in acetonitrile. The flask is subsequently filled to its nominal volume with acetonitrile. The concentration of this solution is 1 g/L.

Working solution of the internal standard:
0.1 mL of the D_8-bisphenol A stock solution are pipetted into a 100 mL volumetric flask and the flask is filled to its nominal value with acetonitrile. The concentration of this solution is 1 mg/L.

The stock and working solutions are stored in sealed vessels at +4 to +6 °C, and they are stable for at least 6 months under these conditions.

The calibration standard solutions are prepared by spiking 2 mL of pooled urine in each case. The dilution error caused by spiking is negligible. The pipetting scheme is shown in Table 3. The standard solutions thus prepared are divided into aliquots in microvials and are stable for at least 1 month when stored sealed at +4 to +6 °C.

2.5 Conditioning the RP 18 columns

The RP 18 cartridges are washed with 6 mL methanol and 6 mL deionised water in that order. Elution is carried out under atmospheric pressure in each case.

Table 3. Pipetting scheme for the preparation of the calibration standard solutions

Volume of the working solution			Volume of the working solution of the ISTD	Volume of pooled urine	Concentration of calibration standard
A [μL]	B [μL]	C [μL]	[μL]	[mL]	[μg/L]
–	–	–	20	2	Blank value
–	–	20	20	2	0.1
–	–	40	20	2	0.2
–	–	100	20	2	0.5
–	20	–	20	2	1
–	40	–	20	2	2
–	100	–	20	2	5
20	–	–	20	2	10
40	–	–	20	2	20
100	–	–	20	2	50

3 Specimen collection and sample preparation

The urine that has been collected in glass vessels (previously rinsed with methanol, see Section 9.4) is analysed immediately or stored at –20 °C and is stable for at least 6 months in this case.

3.1 Sample preparation

The problem of contamination of the sample material with bisphenol A must be taken into consideration in the analytical determination of bisphenol A in urine samples. Sample vessels and instruments that come in contact with the biological material in the laboratory must be thoroughly rinsed with e.g. methanol. Reagent blank values are of great importance in this analysis (see Section 9.4).
The urine is allowed to thaw and reach room temperature overnight. The urine is shaken intensively and then a 2 mL aliquot is transferred to a 10 mL screw-capped test-tube using a pipette. 4 mL of 0.1 M sodium acetate buffer pH 5, 10 μL glucuronidase/arylsulphatase and 20 μL of the working solution of the internal standard are added with pipettes. The test-tube is sealed, shaken briefly and then incubated for at least 2 hours at 37 °C in the incubation cupboard (see Section 9.4). Then 0.1 mL concentrated hydrochloric acid is added and the sample is subsequently thoroughly mixed. The sample is introduced onto the conditioned RP 18 cartridge (see Section 2.5) and passed through it by applying a slight vacuum. Then the cartridge is washed with 3 mL deionised water and subsequently sucked dry for approx. 5 min at about 200 mbar. The analyte is then eluted with 2.0 mL elution solution (acetonitrile/methanol 75:25, v:v) under a slight vacuum into a screw-capped test-tube. Then 4 mL of 1 M sodium hydrogen carbonate solution, 20 μL pentafluorobenzoyl chloride and 0.3 mL iso-octane are added to the eluate. The test-tube is sealed and shaken

vigorously on a shaker for 20 min. The solution is then centrifuged for 10 min at 2200 g. Approx. 200 µL of the organic phase are transferred using a pipette into an autosampler vial, which is sealed tightly. 1 µL of this solution is injected splitless into the GC-MS system. The processed samples are stable at +4 to +6 °C for at least 14 days.

4 Operational parameters

4.1 Operational parameters for gas chromatography

Capillary column:	Material:	Fused silica
	Stationary phase:	HP 1
	Length:	25 m
	Inner diameter:	0.2 mm
	Film thickness:	0.11 µm
Temperatures:	Column:	Starting temperature 80 °C, 1 min isothermal, then increase at a rate of 25 °C/min to 330 °C, then 10 min at the final temperature
	Injector:	300 °C
	Transfer line:	250 °C
Carrier gas:	Helium 5.0 at a constant flow rate of 1.0 mL/min	
Split:	40 mL/min	
Sample volume:	1 µL	

4.2 Operational parameters for mass spectrometry

Ionisation type: Electron impact ionisation (EI)

Ionisation energy: 70 eV

Dwell time: 100 ms

Electron multiplier: Resulting voltage + 400 V

All other parameters must be optimised in accordance with the manufacturer's instructions.

5 Analytical determination

The urine samples processed as described in Section 3.1 are analysed by injecting 1 µL of the iso-octane extract into the gas chromatograph in each case. A quality

Bisphenol A

Table 4. Retention times and recorded masses

Analyte	Retention time [minutes]	Recorded mass [m/z]
Bisphenol A	11.32	601*
		616
D_8-bisphenol A (internal standard)	11.29	609*
		624

The masses marked * are used for quantitative evaluation.

control sample and three reagent blanks (see Section 9.4) consisting of ultrapure water are analysed in each analytical series.

The temporal profiles of the ion traces shown in Table 4 are recorded in the SIM mode.

The retention times shown in Table 4 serve only as a guide. Users of the method must satisfy themselves of the separation power of the capillary column used and the resulting retention behaviour of the substances.

Figure 1 shows an EI mass spectrum of the derivatised bisphenol A. In Figure 2 the GC/MS chromatogram of a processed native urine sample is shown.

6 Calibration

The calibration standard solutions prepared as described in Section 2.4 are processed in the same manner as the urine samples (described in Section 3.1) and analysed by gas chromatography/mass spectrometry as stipulated in Sections 4 and 5. Calibration curves are obtained by plotting the quotients of the peak areas for bisphenol A with that of the internal standard (D_8-bisphenol A) as a function of the concentrations used. The slope of the calibration function and the intercept with the y-axis are calculated by linear regression. If the urine used to prepare the calibration standards exhibits background contamination with bisphenol A, the resulting calibration graph must be shifted in parallel so that it passes through the zero point of the coordinates (the value of the background contamination can be read off from the intercept with the y-axis before the parallel shift). The calibration function is linear in the concentration range between 1.0 µg/L and 50 µg/L urine. Figure 3 shows an example of the linear calibration function for bisphenol A in urine.

7 Calculation of the analytical result

The bisphenol A concentration in urine samples is calculated on the basis of a linear calibration function (cf. Section 6). Quotients are calculated by dividing the peak areas of the analyte by that of the internal standard. These quotients are used to read off the pertinent concentration of bisphenol A in μg per litre urine from the relevant calibration graph. The mean of the reagent blank values that are found must be subtracted from the analytical results for the real samples.

8 Standardisation and quality control

Quality control of the analytical results is carried out as stipulated in the guidelines of the Bundesärztekammer (German Medical Association) [9] and in the special preliminary remarks to this series. A control sample containing a constant concentration of bisphenol A is analysed in order to check the precision of the method. As material for quality control is not commercially available, it must be prepared in the laboratory. For this purpose, a defined quantity of bisphenol A is added to pooled human urine. A six-month supply of this control material is divided into aliquots in previously cleaned 12 mL screw-capped test-tubes and stored in the deep-freezer.
The expected value and the tolerance range of this quality control material are ascertained in a pre-analytical period (one analysis of the control material on each of 20 different days) [10, 11].

9 Evaluation of the method

9.1 Precision

The precision in the series was determined using pooled urine samples that were spiked with two different quantities of bisphenol A, and then processed and analysed as described in the preceding sections. Six replicate assays of the urine samples yielded the precision in the series documented in Table 5.

Table 5. Precision in the series for the assay of bisphenol A in urine (n=6)

Concentration [μg/L urine]	Standard deviation (rel.) [%]	Prognostic range [%]
1	4.1	10.1
10	2.9	7.2

Table 6. Precision from day to day for the assay of bisphenol A in urine (n=6)

Concentration [μg/L urine]	Standard deviation (rel.) [%]	Prognostic range [%]
1	6.7	16.5
10	3.9	9.6

In addition, the precision from day to day was determined. Pooled human urine was spiked with two different quantities of bisphenol A for this purpose. The samples were processed and analysed on 6 different days as described in the previous sections. The resulting precision from day to day is given in Table 6.

9.2 Accuracy

Recovery experiments were carried out at two different concentrations to test the accuracy of the method. The same material that served for the determination of the precision from day to day and the precision in the series was used for this purpose. These urine samples were each processed and analysed 6 times as stipulated in the previous sections. The mean relative recovery rates are shown in Table 7.

Table 7. Mean relative recovery rates for bisphenol A in urine samples (n=6)

n	Concentration [µg/L]	Mean relative recovery [%]	Range [%]
6 (in the series)	1	87	82–92
	10	104	101–109
6 (from day to day)	1	88	77–95
	10	92	87–98

9.3 Detection limits

The detection limit for bisphenol A was calculated as three times the signal/background noise ratio in the temporal vicinity of the analyte signal. The detection limit that may be achieved in theory for bisphenol A by this method is about 0.3 µg/L urine, whereas the quantitation limit is approx. 1.0 µg/L urine. However, these values do not represent the limit of the analytically achievable range, as contamination from the analytical materials hinders the determination of the actual values.

9.4 Sources of error

During development of the method relatively high reagent blank values were initially obtained (up to 5 µg/L). As this may be due to the ubiquitous occurrence of bisphenol A, analysis in the concentration range of relevance to environmental medicine should be performed only after appropriate precautionary measures have been taken. Any contact of the urine sample with plastic must be avoided. Samples should be collected in glass bottles that have been specially cleaned with methanol prior to specimen collection and that can be sealed with either a ground-glass stopper or a

Teflon-coated screw cap. Pasteur pipettes made of glass or Transferpette pipette tips previously cleaned with methanol are suitable for transferring samples in the analytical laboratory. Even the screw-capped test-tubes that are used must be cleaned prior to the start of analysis. Watabe et al. (2004) suggest that the glassware should be heated at 400 °C [12]. In our experience the reagent blank values for this method are generally in the range of 0.1 to 0.2 µg/L, whereby values of 0.5 µg/L may occur in isolated cases. On account of the problem of contamination it is advisable to process three reagent blanks together with the real samples in each analytical series, to distribute the blanks at intervals within the measurement series and to subtract the mean value of the resulting blank values from the analytical results for the real samples.

Different analytical techniques were tested during development of the method. The gentle hydrolysis with glucuronidase/arylsulphatase described here resulted in considerably cleaner analytical samples compared with those obtained by acidic hydrolysis using hydrochloric acid. As distinctly poorer recovery rates were observed for acidic hydrolysis, in our view the method described here using enzymatic hydrolysis is more favourable, when the normal hydrolysis time of 2 hours is observed. The incubation time of 2 hours is also confirmed by the work of Brock et al. (2001) among others [2]. Valuable analytical time can be saved in routine analysis by allowing the hydrolysis to take place overnight.

10 Discussion of the method

The total amount of bisphenol A in urine in the concentration range of relevance to occupational and environmental medicine can be detected by this analytical method. Free bisphenol A is determined in urine by omitting enzymatic hydrolysis. Free bisphenol A in plasma can be assayed as stipulated for the urine method (but without enzymatic hydrolysis).

Very sensitive measurements are achievable due to painstaking cleaning of the glassware together with an easily adaptable analytical method.

Calibration was always carried out in the matrix during the development and validation of the method. It is preferable to perform calibration in this way in order to exclude matrix effects. However, the procedure was also successfully tested using aqueous calibration solutions. As calibration is performed with a labelled internal standard that can compensate for many analytical fluctuations, calibration in water is also possible if there is insufficient pooled urine. However, as the column aged broadening of the peaks became gradually apparent when aqueous standards were used, but this did not occur when calibration was performed in urine. This must be taken into consideration when selecting the type of calibration.

If a high-resolution sector field MS is used, the detection limit that this method may achieve for bisphenol A in theory is about 0.03 µg/L urine and the quantitation limit is approx. 0.1 µg/L urine.

As a derivatisation agent pentafluorobenzoyl chloride fulfils several analytical requirements simultaneously. The high mass of the derivative of m/z 601 (molecular peak: m/z 616 – CH_3) enables interference-free measurement (see Figure 2) when quantification is carried out by GC/MS in the EI mode. In addition, the fluorine atoms in the derivative permit optional use of the GC/MS-NCI technique, which makes even lower detection ranges accessible.

Instruments used:

The analytic measurements were performed on an HP 5890 II+ gas chromatograph coupled with an HP 5970 mass selective detector, both from Agilent.

11 References

[1] *W. Völkel, T. Colnot, G.A. Csanady, J.G. Filser* and *W. Dekant*: Metabolism and kinetics of bisphenol A in humans at low doses following oral administration. Chem. Res. Toxicol. 15, 1281–1287 (2002).
[2] *J.W. Brock, Y. Yoshimura, J.R. Barr, V.L. Maggio, S.R. Graiser, H. Nakazawa* and *L.L. Needham*: Measurement of bisphenol A levels in human urine. J. Expo. Sci. Environ. Epidemiol. 11 (4), 323–328 (2001).
[3] *Deutsche Forschungsgemeinschaft:* List of MAK and BAT Values 2006, 42nd report, Wiley-VCH, Weinheim (2006).
[4] *H. Greim (ed.):* Bisphenol A. Occupational Toxicants – Critical Data Evaluation for MAK Values and Classification of Carcinogens. Vol. 13. Wiley-VCH, Weinheim (1999).
[5] *J. Sajika, K. Takahashi* and *J. Yonekubo*: Sensitive method for the determination of bisphenol A in serum using two systems of high performance chromatography. J. Chrom. B: Biomedical Sciences and Applications 736, 255–261 (1999).
[6] *T. Tsukioka, J. Brock, S. Graiser, J. Nguyen, H. Nakazawa* and *T. Makino*: Determination of trace amounts of bisphenol A in urine by negative-ion chemical-ionization-gas chromatography/mass spectrometry. Anal. Sci. 19, 151–153 (2003).
[7] *H. Imai* and *M. Morita*: National Institute for environmental studies. 4th Annual Meeting of Japan Society of Endocrine Disrupters Research, Tsukuba, Dec. (2001).
[8] *A. Matsumoto, N. Kunugita, K. Kitagawa, T. Isse, T. Oyama, G.L. Foureman, M. Morita* and *T. Kawamoto*: Bisphenol A levels in human urine. Environ. Health Perspect. 111 (1) 101–104 (2003).
[9] *Bundesärztekammer:* Richtlinie der Bundesärztekammer zur Qualitätssicherung quantitativer laboratoriumsmedizinischer Untersuchungen. Dt. Ärztebl. 100, A3335–A3338 (2003).
[10] *J. Angerer* and *G. Lehnert:* Anforderungen an arbeitsmedizinisch-toxikologische Analysen – Stand der Technik. Dt. Ärztebl. 37, C1753–C1760 (1997).
[11] *J. Angerer, Th. Göen* and *G. Lehnert:* Mindestanforderungen an die Qualität von umweltmedizinisch-toxikologischen Analysen. Umweltmed. Forsch. Prax. 3, 307–312 (1998).
[12] *Y. Watabe, T. Kondo, H. Imai, M. Morita, N. Tanaka* and *K. Hosoya*: Reducing bisphenol A contamination from analytical procedures to determine ultralow levels in environmental samples using automated HPLC microanalysis. Anal. Chem. 76, 105–109 (2004).

Authors: *G. Leng, W. Gries*
Examiner: *M. Blaszkewicz*

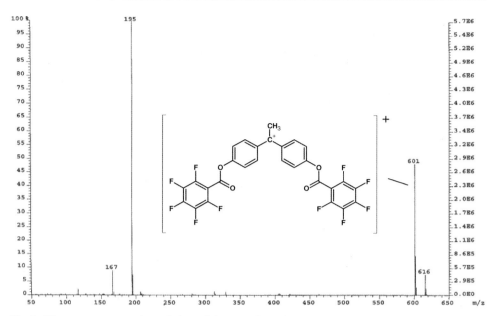

Fig. 1. EI mass spectrum of a solution of the pentafluorobenzoyl chloride derivative of bisphenol A in iso-octane

Bisphenol A

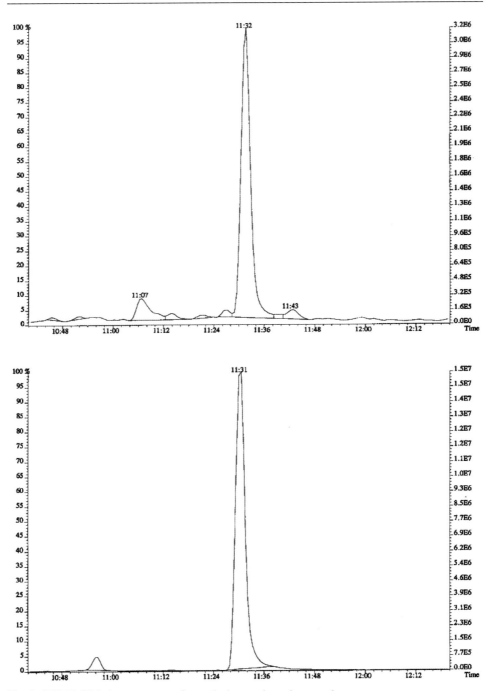

Fig. 2. GC/MS-SIM chromatogram of a worked-up native urine sample

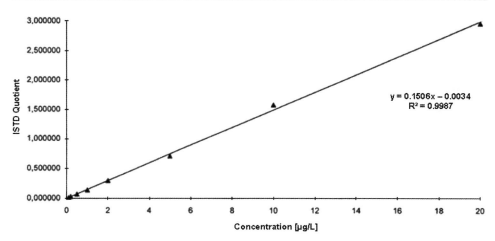

Fig. 3. Linear calibration graph of bisphenol A, prepared in pooled human urine

Di(2-ethylhexyl) phthalate (DEHP) metabolites (2-ethyl-5-hydroxyhexyl phthalate (5OH-MEPH), 2-ethyl-5-oxohexyl phthalate (5oxo-MEHP), mono(2-ethylhexyl) phthalate (MEHP))

Application   Determination in urine

Analytical principle   High performance liquid chromatography/
tandem mass spectrometric detection (LC/MS/MS)

Summary

The method described here permits the assay of two main oxidative metabolites of di(2-ethylhexyl) phthalate (DEHP), i.e. 2-ethyl-5-hydroxyhexyl phthalate (5OH-MEHP) and 2-ethyl-5-oxohexyl phthalate (5oxo-MEHP) as well as the monoester mono(2-ethylhexyl) phthalate (MEHP) in the urine of persons exposed to DEHP at the workplace and in the environment.

After the pH of the urine samples has been adjusted to 6.5 with buffer, D_4-ring-labelled isotope analogues of the target analytes are added and are subjected to enzymatic hydrolysis. The sample processing and the subsequent chromatographic separation are carried out online. The analyte is enriched on a RAM phase and separated from most of the matrix. Then the analyte is transferred to the analytical column by reversing the direction of flow. Detection is performed by means of tandem mass spectrometry with electrospray ionisation. The quantitative determination is achieved by isotope dilution analysis. Aqueous standard solutions, which are processed and analysed in the same manner as the samples, are used for calibration.

The MAK-Collection Part IV: Biomonitoring Methods, Vol. 11.
DFG, Deutsche Forschungsgemeinschaft
Copyright © 2008 WILEY-VCH Verlag GmbH & Co. KGaA, Weinheim
ISBN: 978-3-527-31596-3

2-Ethyl-5-hydroxyhexyl phthalate (5OH-MEHP)

Within-series imprecision: Standard deviation (rel.) s_w=4.3% or 4.9%
Prognostic range u=9.9% or 11.4%
at a concentration of 5.8 µg or 14.4 µg 5OH-MEHP per litre urine and where n=9 determinations

Between-day imprecision: Standard deviation (rel.) s_w=5.3% or 4.5%
Prognostic range u=11.1% or 9.4%
at a concentration of 14.1 or 78.8 µg 5OH-MEHP per litre urine and where n=8 determinations

Accuracy: Recovery rate r=90.9% at 50 µg/L

Quantitation limit: 0.5 µg 5OH-MEHP per litre urine

2-Ethyl-5-oxohexyl phthalate (5oxo-MEHP)

Within-series imprecision: Standard deviation (rel.) s_w=5.9% or 2.5%
Prognostic range u=13.5% or 5.8%
at a concentration of 5.3 µg or 10.3 µg 5oxo-MEHP per litre urine and where n=9 determinations

Between-day imprecision: Standard deviation (rel.) s_w=5.2% or 6.4%
Prognostic range u=13.0% or 13.5%
at a concentration of 11.0 or 55.3 µg 5oxo-MEHP per litre urine and where n=8 determinations

Accuracy: Recovery rate r=84.6% at 50 µg/L

Quantitation limit: 0.5 µg 5oxo-MEHP per litre urine

Mono(2-ethylhexyl) phthalate (MEHP)

Within-series imprecision: Standard deviation (rel.) s_w=9.3% or 9.7%
Prognostic range u=21.4% or 22.4%
at a concentration of 3.2 µg or 2.6 µg MEHP per litre urine and where n=9 determinations

Between-day imprecision: Standard deviation (rel.) s_w=14.5% or 7.9%
Prognostic range u=30.4% or 16.5%
at a concentration of 5.8 or 16.9 µg MEHP per litre urine and where n=8 determinations

Accuracy: Recovery rate r=106% at 50 µg/L

Quantitation limit: 0.5 µg MEHP per litre urine

Di(2-ethylhexyl) phthalate (DEHP)

Di(2-ethylhexyl) phthalate (DEHP)

Mono(2-ethylhexyl) phthalate (MEHP)

Mono(2-ethyl-5-hydroxyhexyl) phthalate (5OH-MEHP, MEHHP)

Mono(2-ethyl-5-oxohexyl) phthalate (5oxo-MEHP, MEOHP)

Di(2-ethylhexyl) phthalate (DEHP) is the most important PVC (polyvinyl chloride) plasticiser used industrially. The DEHP content in PVC can account for 40% of its weight and is even higher in some cases. DEHP is contained in many products such as floor coverings, carpets, cables, pipes, wallpaper, clothing, automobiles etc. [1–5]. The use of phthalate plasticisers is prohibited in toys for children under the age of three years, and meanwhile industry voluntarily avoids using DEHP for food packaging [6, 7]. DEHP production in Germany was approx. 250 000 tonnes in 1995. DEHP is ubiquitously distributed in the environment [8, 9]. The general population presumably takes in DEHP or its hydrolysis product, mono(2-ethylhexyl) phthalate (MEHP), largely by the oral route with food. Inhalative intake is of minor importance. Exposure via inhalation is also probable at certain workplaces in the production and processing of DEHP/PVC [10–13]. DEHP may also enter the body by the intravenous route from medical devices made of plastic such as blood bags, tubes and cannulae [14–18].

In the human body DEHP is metabolised rapidly to MEHP in the first step, and then further to ω-, ω-1-, and β-oxidation products of the side chain. These metabolites are partly bound to glucuronic acid, whereby the degree of glucuronidation of the metabolites in urine may exhibit individual fluctuations of between 0 and 100% [19–23]. Quantitative determination of the metabolites is only possible after enzymatic hydrolysis of the glucuronic acid conjugates [24, 25].

Figure 1 shows the metabolism pathway of DEHP in humans. 23.3% of the absorbed DEHP dose is excreted as 5OH-MEHP in the urine within 24 hours. 15.0% of the dose is excreted as 5oxo-MEHP. MEHP makes up 5.9%. Therefore the three metabo-

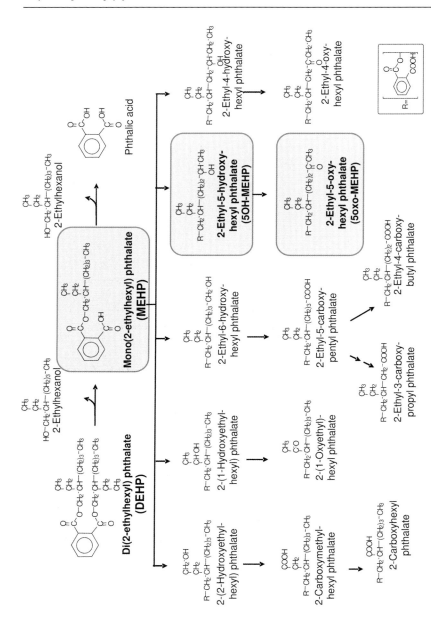

Fig. 1. Metabolism of DEHP in humans (according to [26])

Table 1. MEHP, 5OH-MEHP and 5oxo-MEHP concentrations in the urine of the general population

Source	Country	Group n=(Age)	Median (95th percentile) [µg/L]		
			MEHP	5OH-MEHP	5oxo-MEHP
Koch et al., 2003 [32]	Germany	85 (7 to 63)	10.3 (37.9)	46.8 (224)	36.5 (156)
Becker et al., 2004 [33]	Germany	254 (3 to 14)	7.2 (29.3)	52.1 (185.2)	41.4 (135.6)
Barr et al., 2003 [34]	USA	62	4.5	35.9	28.3
Kato et al., 2004 [23]	USA	127	<DL* (20.4)	15.6 (220)	17.4 (243)

*DL = detection limit.

lites in urine cover 44.2% of the total orally administered DEHP. No difference in the percentages of the excreted metabolites is shown in the dose range from 0.35 to 48.5 mg of DEHP p.o. (equivalent to 5 µg/kg to 700 µg/kg body weight) [22]. Further oxidative metabolites such as 2-ethyl-5-carboxypentyl phthalate (5cx-MEPP) and 2-carboxymethylhexyl phthalate (2cx-MMHP) increase the proportion of the dose excreted in the urine to 67% in 24 hours and 74.3% in 48 hours [22]. Excretion exhibits biphasic to multiphasic kinetics. The half-life is about 5 hours for the excretion of MEHP in the slower second excretion phase, and for 5OH-MEHP and 5oxo-MEHP the half-life is about 10 hours [21, 22].

The secondary metabolites 5OH-MEHP and 5oxo-MEHP are more suitable as parameters of exposure to DEHP than MEHP due to their higher concentrations in urine, their insensitivity to contamination and their longer half-life [27].

The Commission has assigned DEHP to Carcinogen Category 4. The MAK value (maximum permissible concentration at the workplace) for DEHP is 10 mg/m^3 [12, 28]. A comprehensive review of the toxicological aspects of DEHP can be found in the MAK value documentation [12]. To date no BAT value has been assigned. Preventive limit values for the daily intake of DEHP have been stipulated by various authorities. These include 37 µg/kg body weight/day (TDI: tolerable daily intake) set by the EU Scientific Committee on Toxicity, Ecotoxicity and the Environment, 50 µg/kg body weight/day by EFSA (European Food Safety Authority) and 20 µg/kg body weight/day (RfD: reference dose) by the US EPA (Environmental Protection Agency) [29–31].

Table 1 shows an overview of the studies carried out to assess the internal exposure of the general population to DEHP.

Maximum values for the secondary metabolites were measured in the mg/L range in all the studies. The concentration range to be covered by the method is between 0.5 and 2000 µg/L of the metabolites in urine.

Authors: *H. M. Koch, J. Angerer*
Examiners: *N. Bittner, E. R. Trösken, W. Völkel*

Di(2-ethylhexyl) phthalate (DEHP) metabolites (2-ethyl-5-hydroxyhexyl phthalate (5OH-MEPH), 2-ethyl-5-oxohexyl phthalate (5oxo-MEHP), mono(2-ethylhexyl) phthalate (MEHP))

Application	Determination in urine
Analytical principle	High performance liquid chromatography/ tandem mass spectrometric detection (LC/MS/MS)

Contents

1 General principles
2 Equipment, chemicals and solutions
2.1 Equipment
2.2 Chemicals
2.3 Solutions
2.4 Calibration standards
2.4.1 Internal standard (ISTD)
2.4.2 Calibration standards
3 Specimen collection and sample preparation
3.1 Sample preparation
4 Operational parameters
4.1 Operational parameters for high performance liquid chromatography
4.2 Operational parameters for mass spectrometry
5 Analytical determination
6 Calibration
7 Calculation of the analytical result
8 Standardisation and quality control
9 Evaluation of the method
9.1 Precision
9.2 Accuracy

The MAK-Collection Part IV: Biomonitoring Methods, Vol. 11.
DFG, Deutsche Forschungsgemeinschaft
Copyright © 2008 WILEY-VCH Verlag GmbH & Co. KGaA, Weinheim
ISBN: 978-3-527-31596-3

9.3 Detection limit
9.4 Sources of error
10 Discussion of the method
11 References

1 General principles

After the pH of the urine samples has been adjusted to 6.5 with buffer, D_4-ring-labelled isotope analogues of the target analytes are added and subjected to enzymatic hydrolysis. The sample processing and the subsequent chromatographic separation are carried out online. The analyte is enriched on a RAM phase and separated from most of the matrix. Then the analyte is transferred to the analytical column by reversing the direction of flow. Detection is performed by means of tandem mass spectrometry with electrospray ionisation. The quantitative determination is achieved by isotope dilution analysis. Aqueous standard solutions, which are processed and analysed in the same manner as the samples, are used for calibration.

2 Equipment, chemicals and solutions

2.1 Equipment

LC/MS/MS system comprising a quaternary pump system with injection system up to 900 µL sample volume and an isocratic pump, a 6-port valve, and a tandem mass spectrometric detector with a turbo ion spray interface and a PC system for data evaluation.

HPLC columns:
RAM phase (Restricted Access Material): LiChrospher® RP-8 ADS, length 25 mm; inner diameter: 4 mm; particle diameter: 25 µm (e.g. Merck No. 1.50209.0001)
Analytical column: Luna Phenyl-Hexyl, length: 150 mm; inner diameter: 4.6 mm; particle diameter: 3 µm (e.g. Phenomenex No. 00F-4256-E0)
Precolumn: Phenylpropyl, 4×3 mm (e.g. Phenomenex No. AJO-4351)
Particle filter (e.g. Supelco No. 57677)

250 mL Urine containers (sealable plastic bottles e.g. Sarstedt No. 77.577)

25, 50, 100, 200, 250 and 1000 mL Volumetric flasks

1.8 mL Screw-capped glass vials and screw caps with PTFE septa

Variably adjustable pipettes with 10, 100, 500 and 2000 µL pipette tips (e.g. Multipipettes from Eppendorf)

400 mL Glass beaker

1000 mL Measuring cylinder

Glass Pasteur pipettes

Drying cupboard

pH test strips (e.g. Macherey-Nagel Pehanon 5.2–6.8 No. 904 16)

Magnetic stirrer with stirring rod

2.2 Chemicals

β-Glucuronidase, E. coli K12 (e.g. Roche Biomedical No. 03707 598001)

Acetonitrile LiChrosolv® (e.g. Merck No. 1.14291.2500)

Glacial acetic acid, p.a. (e.g. Merck No. 1.00063.2500)

Methanol SupraSolv® (e.g. Merck No. 1.06011.2500)

Ammonium acetate p.a. (e.g. Merck No. 1.01116.0500)

Ultrapure water (equivalent to ASTM Type 1)

Pronase E (e.g. Merck No. 1.07433.0001)

2-Ethyl-5-hydroxyhexyl phthalate (e.g. Cambridge Isotope Laboratories, No. ULM-4662-1.2)

2-Ethyl-5-oxohexyl phthalate (e.g. from Institut für Dünnschichttechnologie und Mikrosensorik e.V. (IDM))

D_4-analogues of 5oxo-MEHP and 5OH-MEHP (e.g. Institut für Dünnschichttechnologie und Mikrosensorik e.V. (IDM))

Mono(2-ethylhexyl) phthalate (MEHP) (e.g. Cambridge Isotope Laboratories, No. ULM-4583-1.2)

D_4-mono(2-ethylhexyl) phthalate (e.g. synthesis according to Dirven et al. 1993 [35])

Mono-n-octylphthalate (MOP) (e.g. Cambridge Isotope Laboratories, No. ULM-4593-1.2)

The calibration standards had a chemical purity of >95%. The purity of the isotopes of each internal D_4-standard was checked by means of LC/MS/MS. They contained no detectable non-labelled or partially labelled (D_3- or D_2-) compounds.

2.3 Solutions

1 M Ammonium acetate buffer (pH 6.5):
19.3 g ammonium acetate are weighed in a 250 mL volumetric flask. The volumetric flask is subsequently filled to its nominal volume with ultrapure water. After the solution has been transferred to a 400 mL glass beaker, the pH is adjusted to 6.5

(checked with the aid of pH test strips) by adding glacial acetic acid drop by drop while the contents are stirred constantly (magnetic stirrer with stirring rod). This buffer is stable for 8 weeks in the refrigerator.

Eluent (RAM):
940 mL ultrapure water, 10 mL glacial acetic acid and 50 mL methanol are measured into a 1000 mL measuring cylinder. The eluent is mixed thoroughly. No degassing or further adjustments are necessary. This eluent serves only to transfer the analytes onto the RAM phase.

Eluent A:
100 mL acetonitrile and 10 mL glacial acetic acid are placed in a 1000 mL volumetric flask and it is filled to its nominal volume with water. The eluent is mixed thoroughly.

Eluent B:
900 mL acetonitrile and 10 mL glacial acetic acid are placed in a 1000 mL volumetric flask and it is filled to the nominal volume with water. The eluent is mixed thoroughly.

Eluent C:
Methanol SupraSolv®.

2.4 Calibration standards

2.4.1 Internal standard (ISTD)

Starting solution of the ISTD:
Approx. 5.0 mg of each internal standard, D_4-MEHP, D_4-5OH-MEHP and D_4-5oxo-MEHP, are weighed exactly and placed into a 25 mL volumetric flask. The flask is then filled to its nominal volume with acetonitrile (200 mg/L).
This starting solution is stored at $-18\,°C$ in a screw-capped glass vial with a Teflon seal and it is stable for at least one year under these conditions.

Spiking solution of the ISTD:
1 mL of the starting solution of the ISTD is pipetted into a 100 mL volumetric flask. The volumetric flask is subsequently filled to its nominal volume with ultrapure water (2 mg/L).
The spiking solution of the ISTD can be kept in the refrigerator at $+4\,°C$ for at least 6 weeks.

2.4.2 Calibration standards

Starting solution:
Approx. 10.0 mg each of MEHP, 5OH-MEHP and 5oxo-MEHP, weighed exactly, are placed into a 25 mL volumetric flask. The flask is then filled to its nominal volume with acetonitrile (400 mg/L).
This starting solution is stored at −18 °C in a screw-capped glass vial with a Teflon seal. It is stable for at least one year under these conditions.

Working solution A:
1.0 mL of the starting solution is placed into a 200 mL volumetric flask. The volumetric flask is subsequently filled to its nominal volume with ultrapure water (2 mg/L).

Working solution B:
20.0 mL of working solution A are placed in a 200 mL volumetric flask. The flask is filled to its nominal volume with ultrapure water (200 µg/L).

Working solution C:
5.0 mL of working solution B are placed in a 250 mL volumetric flask. The flask is filled to its nominal volume with ultrapure water (4 µg/L).
The working solutions can be kept in the refrigerator at +4 °C for at least 6 weeks.
The calibration standard solutions for the determination of the DEHP metabolites in urine are prepared in water. Working solutions A, B and C are diluted with water to a volume of 100 mL in a 100 mL volumetric flask as stipulated in the pipetting scheme in Table 2. The calibration standard solutions are processed and analysed in the same manner as the urine samples as described in Sections 3 and 4.

Table 2. Pipetting scheme for the preparation of the calibration standard solutions

Volume of working solutions			Ultrapure water [mL]	Final volume of the calibration standard solution [mL]	Concentration of the calibration standard solution [µg/L]
A [mL]	B [mL]	C [mL]			
100*	–	–	–	100	2000
25	–	–	75	100	500
–	25	–	75	100	50
–	10	–	90	100	20
–	5	–	95	100	10
–	–	25	75	100	1
–	–	10	90	100	0.4

* Working solution A serves without further dilution as a calibration standard solution with a concentration of 2000 µg/L.

3 Specimen collection and sample preparation

The urine is collected in sealable 250 mL plastic bottles and stored in the deep-freezer at approx. −18 °C until processing. The urine can be stored in this manner for at least one year.

3.1 Sample preparation

The urine is allowed to thaw overnight. When the urine samples have reached room temperature, they are shaken vigorously. Aliquots of 1 mL are transferred to 1.8 mL screw-capped glass vials. 200 µL of 1M ammonium acetate buffer (pH 6.5), 50 µL of the spiking solution of the ISTD and 5 µL of β-glucuronidase are added using a pipette, the vials are sealed and then shaken gently. The samples are warmed at 37 °C in the drying cupboard for 1.5 h in order to hydrolyse the glucuronides of the phthalate metabolites. After the samples have cooled, they are centrifuged for 15 min at 3500 g. The liquid phase is separated from the precipitate using a glass Pasteur pipette and then transferred to a new 1.8 mL screw-capped glass vial. A volume of 600 µL of the hydrolysed sample is injected into the LC/MS/MS system for analysis.

4 Operational parameters

The analytical measurement is performed by a combination of instruments consisting of a gradient pump, an isocratic pump and a LC/MS/MS system capable of negative electrospray ionisation.

4.1 Operational parameters for high performance liquid chromatography

Enrichment column:	Material:	Steel
	Length:	25 mm
	Inner diameter:	4 mm
	Column packing:	Restricted Access Material (RAM); Merck RP-8 ADS, 25 µm
Separation column:	Material:	Steel
	Length:	150 mm
	Inner diameter:	4.6 mm
	Column packing:	Phenomenex Luna Phenyl-Hexyl, 3 µm
Separation principle:	Reversed phase	
Temperature:	Room temperature	

Detection:	Tandem mass spectrometric detector	
Mobile phase:	Eluent (RAM):	1% glacial acetic acid 5% methanol 94% water (v/v/v)
	Eluent A:	1% glacial acetic acid 10% acetonitrile 89% water (v/v/v)
	Eluent B:	1% glacial acetic acid 90% acetonitrile 9% water (v/v/v)
	Eluent C:	Methanol
Gradient:	See Table 3	

Table 3. Gradient program for chromatographic separation

Time [min]	Eluent A [%]	Eluent B [%]	Eluent C [%]
0.0	45	45	10
10.1	12.5	77.5	10
13.0	10	85	5
13.5	0	100	0
16.0	0	0	100
21.1	0	100	0
24.4	45	45	10

Stop time:	25 min
Flow rate:	Gradient pump: 0.8 mL/min Isocratic pump: 1.5 mL/min
Injection volume:	600 µL
Valve settings:	See Table 4

All other parameters must be optimised in accordance with the manufacturer's instructions.

Table 4. Program for the valve settings

Time [min]	Valve position	Analytical step
0.0	A	RAM loading
6.0	B	Transfer
8.0	A	Chromatographic separation
15.6	B	Washing
21.0	A	Reconditioning

4.2 Operational parameters for mass spectrometry

Settings for the ion source:

Ionisation mode:	ESI negative
Electrospray nozzle voltage:	–4000 V
Nebulizer Gas:	Nitrogen; 35 psi
Turbo Heater Gas:	Nitrogen; 500 °C; 68 psi
Curtain Gas:	Nitrogen; 45 psi

Settings for the analyser:

Q1 Resolution:	unit
Q3 Resolution:	low
Settling Time:	5 msec
MR Pause:	5 msec
Collision gas:	Nitrogen; 4 instrument units
Scan Time:	75 msec
Analyte-specific parameters:	See Table 5

All the ion source and MRM parameters are specific for each instrument and must be individually set by the user by means of the appropriate calibration routines of the MS/MS system.

Table 5. Analyte-specific parameters

Parameter	Parent ion	Daughter ion	DP	FP	EP	CE
5-OH-MEHP	293	77	–16	–330	–8.5	–40
	293	121	–16	–330	–8.5	–24
D4-5OH-MEHP	297	125	–21	–340	–11.5	–26
5oxo-MEHP	291	77	–11	–340	–9	–40
	291	121	–16	–310	–9	–20
D4-5oxo-MEHP	295	125	–31	–350	–11.5	–26
MEHP	277	127	–21	–350	–9.5	–26
	277	134	–21	–350	–9.5	–26
D4-MEHP	281	138	–21	–340	–10	–20
MOP	277	127	–21	–350	–9.5	–26
	277	134	–21	–350	–9.5	–26

DP = declustering potential [V], FP = focussing potential [V], EP = entrance potential [V], CE = collision energy [V]

The measurement conditions listed here were established for the configuration of instruments used in this case and they must be optimised for other instruments in accordance with the manufacturer's instructions.

5 Analytical determination

Using the isocratic pump and the eluent (RAM), 600 µL of the urine samples processed as described in Section 3.1 are introduced onto the RAM phase at valve position A (Figure 2). The flow rate is 1.5 mL/min in this case.
After this clean-up and enrichment step, the analytes are transferred from the RAM phase to the analytical column at valve position B, reversal of the solvent flow and increase in the organic proportion (see Figure 3).
If the measured values are above the range of the calibration graphs (>2000 µg/L), the urine samples are appropriately diluted (1:1, 1:5 or 1:10) with ammonium acetate buffer (1M, pH 6.5), processed anew and measured. Two quality control samples (high concentration level and low concentration level) and a reagent blank are analysed in each analytical series. Ultrapure water serves as a blank and is subjected to the processing described above instead of urine.

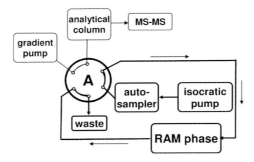

Fig. 2. Switch circuit at valve position A – loading the RAM phase

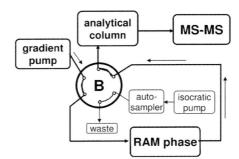

Fig. 3. Switch circuit at valve position B – analytical separation

Table 6. Retention times and detected ion transitions

Analyte	Retention time [min]	Ion transitions (MS/MS, ESI neg. mode)	
		Q 1	Q 3
5-OH-MEHP	12.30	293	77
			121*
D_4-5OH-MEHP	12.25	297	125*
5oxo-MEHP	13.05	291	77*
			121
D_4-5oxo-MEHP	13.00	295	125*
MEHP	15.18	277	127
			134*
D_4-MEHP	15.15	281	138*
MOP	15.37	277	127*
			134

The masses marked * are used for quantitative evaluation.

The temporal traces of the ion transitions in Table 6 are recorded in the MRM mode of the tandem mass spectrometer (ESI negative mode).
The retention times shown in Table 6 serve only as a guide. Users of the method must satisfy themselves of the separation power of the capillary column used and the resulting retention behaviour of the substances. The chromatogram of a native urine sample is shown in Figure 4.

6 Calibration

The calibration standard solutions (Section 2.4) are processed in the same manner as the urine samples (Section 3.1) and analysed by liquid chromatography/mass spectrometry as described in Section 4. Calibration curves are obtained by plotting the quotients of the peak areas of the calibration standards and of the appropriate internal standard as a function of the concentrations used. It is not necessary to plot a complete calibration graph for every analytical series. It is sufficient to measure a calibration standard solution in the intermediate concentration range (20 µg/L to 200 µg/L) in each analytical series. The ratio of the result obtained for this standard and the result for the equivalent standard in the complete calibration graph is calculated. Using this quotient, each result read off the calibration graph is corrected (one-point calibration). It is essential to take the analytical blank value into account for the MEHP metabolite (i.e. calibration through the zero point of the coordinates).
New calibration graphs should be plotted if the quality control results indicate systematic deviation.

The calibration curve is linear between the detection limit and 2000 µg of the DEHP metabolites per litre urine.

7 Calculation of the analytical result

The quotients calculated from the peak areas of the analyte and of the internal standard are used to read off the pertinent concentration of the metabolites in µg per litre urine from the relevant linear calibration graph.
A reagent blank value can occur for MEHP in the vicinity of the quantitation limit (0.5 µg/L). The mean blank value plus 3 times its standard deviation must be subtracted from the result for MEHP to compensate for this contamination. No blank values are to be expected in the case of the secondary metabolites.

8 Standardisation and quality control

Quality control of the analytical results is carried out as stipulated in the guidelines of the Bundesärztekammer (German Medical Association) [36, 37] and in the special preliminary remarks to this series.
Standard solutions of all the metabolites were continually injected with the injection pump of the LC/MS/MS system in order to establish the MS/MS conditions. Two specific MS/MS transitions were determined and calibrated for each analyte. The corresponding molecule ion peaks were selected as parent ions in each case. Quantification was based on the main fragmentation, the second trace was evaluated to assure the result (qualifier). The fragment pattern of MEHP, 5OH-MEHP, 5oxo-MEHP and a deuterium-labelled standard D_4-5OH-MEHP (as an example) are shown in Figure 5.
Urine control samples containing constant native concentrations of the DEHP metabolites are used to check the precision. As material for quality control is not commercially available, it must be prepared in the laboratory. For this purpose individual urine samples are pooled in such a manner that one pooled sample contains a low concentration and the other a high concentration of the DEHP metabolites. A six-month supply of this control material is prepared, divided into aliquots in screw-capped jars with a Teflon septum and stored in the deep-freezer. The concentration of this control material should lie within the decisive concentration range. The expected value and the tolerance range of this quality control material are ascertained in a pre-analytical period (one analysis of the control material on each of 20 different days) [38, 39]. As the metabolites are present in both the free and the conjugated form in these native control urine samples, this method of quality assurance also represents a check that hydrolysis of the conjugates is complete.

9 Evaluation of the method

9.1 Precision

Two control urine sample containing different levels of the DEHP metabolites were used to determine the precision in the series. Nine analyses of these urine samples yielded the precision in the series documented in Table 7.
In addition, the precision from day to day was determined. Two control urine samples were also used for this purpose. These samples were processed and analysed on each of 8 different days. The precision results are shown in Table 8.

Table 7. Precision in the series for the determination of 5OH-MEHP, 5oxo-MEHP and MEHP in two native control urine samples (n=9)

Parameter	Native control urine samples	Low concentration	High concentration
	Creatinine concentration [g/L]	0.32	1.18
5OH-MEHP	Concentration [µg/L]	5.8	14.4
	Standard deviation (rel.) s_w [%]	4.3	4.9
	Prognostic range u [%]	9.9	11.4
5oxo-MEHP	Concentration [µg/L]	5.3	10.3
	Standard deviation (rel.) s_w [%]	5.9	2.5
	Prognostic range u [%]	13.5	5.8
MEHP	Concentration [µg/L]	3.2	2.6
	Standard deviation (rel.) s_w [%]	9.3	9.7
	Prognostic range u [%]	21.4	22.4

Table 8. Precision from day to day for the determination of 5OH-MEHP, 5oxo-MEHP and MEHP in two native control urine samples (n=8)

Parameter	Native control urine samples	Low concentration	High concentration
	Creatinine concentration [g/L]	0.65	1.32
5OH-MEHP	Concentration [µg/L]	14.1	78.8
	Standard deviation (rel.) s_w [%]	5.3	4.5
	Prognostic range u [%]	11.1	9.4
5oxo-MEHP	Concentration [µg/L]	11.0	55.3
	Standard deviation (rel.) s_w [%]	5.2	6.4
	Prognostic range u [%]	16.0	13.5
MEHP	Concentration [µg/L]	5.8	16.9
	Standard deviation (rel.) s_w [%]	14.5	7.9
	Prognostic range u [%]	30.4	16.5

9.2 Accuracy

Recovery experiments were performed to check the accuracy of the method. For this purpose 8 individual urine samples without the analytes or spiked with defined quantities of the DEHP metabolites were analysed. The creatinine levels of the individual urine samples were between 0.24 g/L and 2.12 g/L. The relative recovery rates are summarised in Table 9.

Table 9. Relative recovery rates for 5OH-MEHP, 5oxo-MEHP and MEHP in spiked individual urine samples (n = 8)

Parameter	Native concentration [µg/L]	Spiked concentration [µg/L]	Relative recovery r [%]	Range [%]
5OH-MEHP	0.7–147.9	50	90.9	84.8–95.0
5oxo-MEHP	0.5–93.1	50	84.6	76.9–101
MEHP	<QL–42.7	50	106	97.3–110

QL = Quantitation limit

9.3 Detection limit

The detection limit of the three DEHP metabolites was approx. 0.25 µg per litre urine under the conditions given here. The detection limit was estimated as three times the signal/background noise ratio in the temporal vicinity of the analyte signal. The quantitation limit was about 0.5 µg per litre urine and was estimated as six times the signal/background noise ratio.

9.4 Sources of error

The laboratory equipment used should be tested to ensure it is free of DEHP and MEHP in order to avoid high blank values for MEHP. It is extremely important to test the sample vessels used to store the urine and standard samples. Nowadays DEHP and MEHP have become ubiquitous chemicals in the environment, and are also present in the laboratory. In the case of high DEHP contamination MEHP can be released from DEHP during processing, storage or measurement and thus falsify the analytical result. In contrast, the determination of the secondary metabolites 5OH-MEHP and 5oxo-MEHP is not susceptible to contamination.

For the chromatographic separation it is essential to ensure separation of MEHP from its structural isomers MOP (mono-n-octyl phthalate), as differentiation proves impossible due to their fragmentation spectra with identical fragment masses. In the mass traces of the fragment masses of the secondary metabolites other peaks are visible (in low concentrations however) in addition to the peaks for 5OH-MEHP and

5oxo-MEHP. These peaks are presumably metabolites of DEHP/MEHP oxidised at other positions of the side chain. The 5OH-MEHP and 5oxo-MEHP peaks must be separated from these other peaks by chromatography, as these isomers exhibit the same fragment pattern.

The direct introduction of urine onto the RAM phase proved extremely robust. A pre-column should be used to prolong the useful life of the analytical column. Furthermore, a 5 μm particle filter should be fitted between the autosampler and the 6-port valve in order to protect the RAM phase from particle accumulation. In some cases elevation of the pressure of the RAM phase is due to precipitation of protein. This can be prevented by longer washing times (up to 10 min at a flow rate of 1.5 mL/min) after introduction of the sample onto the RAM phase. Clogged RAM phases can be cleaned again by rinsing with Pronase E solution (0.5% in ammonium acetate buffer pH 6.5).

10 Discussion of the method

The method described here permits reliable and accurate determination of primary and secondary DEHP metabolites at the concentration range due to environmental exposure. Adaptation to the range of interest to occupational medicine is possible. The reliability criteria are considered very good due to the isotope-labelled internal standards, and the examiners of the method were able to replicate them with no problems. However, the examiners point out that the use of an LC/MS/MS instrument of higher sensitivity than that used in this case can lead to limitation of the linear range (examiners: 0.4 to 500 μg/L).

On account of the susceptibility to contamination in the determination of MEHP, which can be kept under control to some extent in the laboratory but not during specimen collection, the secondary metabolites of DEHP are regarded as the parameters of choice.

Commercially available $^{13}C_4$-MEHP (ring-1,2-$^{13}C_2$, dicarboxyl-$^{13}C_2$) can serve as an alternative internal standard for MEHP (e.g. Cambridge Isotope Laboratories, No. CLM-4584-1.2).

In addition to 5OH-MEHP and 5oxo-MEHP, further secondary metabolites such as 5cx-MEHP [40] can be included in the method. Moreover, monoester metabolites (with known contamination risks) of other phthalates such as mono-n-butyl phthalate, mono-iso-butyl phthalate, monobenzyl phthalate and mono-n-octyl phthalate [25] may also be analysed.

Instruments used:
LC/MS/MS system consisting of a Hewlett-Packard HP 1100 with autosampler, quaternary pump and device for degassing the eluents, equipped with an injection system up to 900 μL sample volume and a Merck-Hitachi L6000A isocratic pump, an API 2000 tandem mass spectrometric detector from Applied Biosystems with a 10-port valve (only 6 ports were used) and turbo ionspray interface. "Analyst" software from Applied Biosystems was used for evaluation.

11 References

[1] *IARC International Agency for Research on Cancer:* Some industrial chemicals and dyestuffs. Vol. 29. In: IARC Monographs on the Evaluation of the Carcinogenic Risk of Chemicals to Humans, 416 (1982).
[2] *IARC International Agency for Research on Cancer:* Di(2-ethylhexyl)phthalate. In: IARC Monographs 77, 41–148 (2000).
[3] *H. Fromme:* Organische Stoffe – Phthalate. In: Praktische Umweltmedizin: Chemische Faktoren, Part 4 (Beyer and Eis, eds.). Springer Verlag, Berlin (1999).
[4] *G. Rippen:* Di(2-ethylhexyl)phthalat. In: Handbuch Umweltchemikalien, 51st supplement. Ecomed., Landsberg (2000).
[5] *C. Böhme:* Chemikalien mit östrogenem Potential in Lebensmitteln und kosmetischen Mitteln. Bundesgesundheitsblatt 41, 340–343 (1998).
[6] *Commission of the European Communities:* Commission decision of 7 December 1999 adopting measures prohibiting the placing on the market of toys and childcare articles intended to be placed in the mouth by children under three years of age made of soft PVC containing one or more of the substances di-iso-nonyl phthalate (DINP), di(2-ethylhexyl) phthalate (DEHP), dibutyl phthalate (DBP), di-iso-decyl phthalate (DIDP), di-n-octyl phthalate (DNOP), and butylbenzyl phthalate (BBP), (notified under document number C (1999) 4836).
[7] *Commission of the European Communities:* Commission decision of 20 February 2004 amending Decision 1999/815/EC concerning measures prohibiting the placing on the market of toys and childcare articles intended to be placed in the mouth by children under three years of age made of soft PVC containing certain phthalates (Text with EEA relevance) (notified under document number C (2004) 524).
[8] *A. Leisewitz:* Endokrine Stoffe. In: Abwassertechnische Vereinigung (ed.): ATV-Schriftenreihe Vol. 15, 22–37. Hennef (1999).
[9] *P.M. Lorz, F.K. Towae, W. Enke, R. Jäckh and N. Bhargava:* Phthalic acid and derivatives. In: Ullmann's Encyclopedia of Industrial Chemistry. Wiley-VCH, Weinheim (2002).
[10] *BUA Beratergremium für umweltrelevante Altstoffe:* Di-(2-ethylhexyl)phthalat. BUA-Stoffbericht 4, VCH-Verlagsgesellschaft, Weinheim (1986).
[11] *BUA Beratergremium für umweltrelevante Altstoffe:* Ergänzungsberichte I: Di-(2-ethylhexyl)phthalat (Nr. 4), BUA-Stoffbericht 114, VCH-Verlagsgesellschaft, Weinheim (1993).
[12] *H. Greim (ed.):* Di(2-ethylhexyl)phthalat (DEHP). Gesundheitsschädliche Arbeitsstoffe. Toxikologisch-arbeitsmedizinische Begründungen von MAK-Werten. Issue 35. Wiley-VCH (2002).
[13] *R. Kavlock, K. Boeckelheide, R. Chapin, M. Cunningham, E. Faustman* and *P. Foster:* NTP Center for the Evaluation of Risks to Human Reproduction: phthalates expert panel report on the reproductive and developmental toxicity of di(2-ethylhexyl)phthalate. Reprod. Toxicol. 16, 529–653 (2002).
[14] *H. Planck, (ed.):* Kunststoffe und Elastomere in der Medizin. Kohlhammer, Stuttgart (1993).
[15] *A. Bruder, S. Lindner, J. Mügge, R. Saffert* and *E. Spindler:* PVC – der Werkstoff für Medizinprodukte. Swiss Plastics 21, 5–10 (1999).
[16] *FDA U.S. Food and Drug Administration:* Safety assessment of di(2-ethylhexyl)phthalate (DEHP) released from medical devices. Center for Devices and Radiological Health, U.S. Food and Drug Administration, 12709 Twinbrook Parkway. Rockville, MD 20852, USA (2001).
[17] *H.M. Koch, J. Angerer, H. Drexler, R. Eckstein* and *V. Weisbach:* Di(2-ethylhexyl)phthalate (DEHP) exposure of voluntary plasma and platelet donors. Int. J. Hyg. Environ. Health 208(6), 489–498 (2005).
[18] *V. Weisbach, H.M. Koch, J. Angerer* and *R. Eckstein:* Di(2-ethylhexyl)phthalate (DEHP) exposure of apheresis donors is procedure-related (Letter). Transfusion 46(8) 1457–1458 (2006).
[19] *C.C. Peck* and *P.W. Albro:* Toxic potential of the plasticizer di(2-ethylhexyl) phthalate in the context of its disposition and metabolism in primates and man. Environ. Health Perspect. 45, 11–17 (1982).
[20] *P. Schmid* and *C. Schlatter:* Excretion and metabolism of di(2-ethylhexyl)phthalate in man. Xenobiotica 15, 251–256 (1985).

[21] *H.M. Koch, H.M. Bolt* and *J. Angerer:* Di(2-ethylhexyl)phthalate (DEHP) Metabolites in Human Urine and Serum After a Single Oral Dose of Deuterium Labelled DEHP. Arch. Toxicol. 78, 123–130 (2004).
[22] *H.M. Koch, H.M. Bolt, R. Preuss* and *J. Angerer:* New Metabolites of Di-(2-ethylhexyl)phthalate (DEHP) in Human Urine and Serum After Single Oral Doses of Deuterium Labelled DEHP. Archives of Toxicology 79(7), 367–376 (2005).
[23] *K. Kato, M.J. Silva, J.A. Reidy, D. Hurtz III, N.A. Malek, L.L. Needham, H. Nakazawa, D.B. Barr* and *A.M. Calafat:* Mono(2-Ethyl-5-Hydroxyhexyl) Phthalate and Mono(2-Ethyl-5-Oxohexyl) Phthalate as Biomarkers for Human Exposure Assessment to Di(2-Ethylhexyl) Phthalate. Environ. Health Perspect. 112, 327–330 (2004).
[24] *B.C. Blount, K.E. Milgram, M.J. Silva, N.A. Malek, J.A. Reidy, L.L. Needham* and *J.W. Brock:* Quantitative detection of eight phthalate metabolites in human urine using HPLC-APCI-MS/MS. Anal. Chem. 72, 4127–4134 (2000).
[25] *H.M. Koch, L.M. Gonzalez-Reche* and *J. Angerer:* On-line cleanup by multidimensional LC-ESI-MS/MS for high throughput quantification of primary and secondary phthalate metabolites in human urine. J Chromatogr. B 784, 169–182 (2003).
[26] *H.M. Koch, H. Drexler* and *J. Angerer:* Die innere Belastung der Allgemeinbevölkerung mit Di(2-ethylhexyl)phthalat (DEHP). Umweltmed. Forsch. Prax. 8(1), 15–23 (2003).
[27] *CSTEE Scientific Committee for Toxicity, Ecotoxicity and the Environment:* Opinion on the results of a second Risk Assessment of: BIS(2-ETHYLHEXYL) PHTHALATE [DEHP] HUMAN HEALTH PART. Adopted by the CSTEE during the 41th plenary meeting of 8 January 2004. http://europa.eu.int/comm/health/ph_risk/committees/sct/documents/out214_en.pdf
[28] *Deutsche Forschungsgemeinschaft:* List of MAK and BAT Values 2006, 42nd report, Wiley-VCH, Weinheim (2006).
[29] *CSTEE Scientific Committee for Toxicity, Ecotoxicity and the Environment:* Opinion on Phthalate migration from soft PVC toys and child-care articles – opinion expressed at the 6th CSTEE plenary meeting, Brussels, 26/27 November 1998 http://europa.eu.int/comm/food/fs/sc/sct/out19_en.html
[30] *EPA U.S. Environmental Protection Agency:* Integrated Risk Information System (IRIS) on Di(2-ethylhexyl)phthalate. National Center for Environmental Assessment, Office of Research and Development, Washington, DC. (1999).
[31] *EFSA European Food Safety Authority:* Opinion of the Scientific Panel on Food Additives, Flavourings, Processing Aids and Materials in Contact with Food (AFC) on a request from the Commission related to bis(2-ethylhexyl)phthalate (DEHP) for use in food contact materials. The EFSA Journal 243, 1–20 (2005).
[32] *H.M. Koch, B. Rossbach, H. Drexler* and *J. Angerer:* Internal exposure of the general population to DEHP and other phthalates – determination of secondary and primary phthalate monoester metabolites in urine. Environ. Res. 93(2), 177–185 (2003)
[33] *K. Becker, M. Seiwert, J. Angerer, W. Heger, H.M. Koch, R. Nagorka, E. Roßkamp, C. Schlüter, B. Seifert* and *D. Ullrich:* DEHP metabolites in urine of children and DEHP in house dust. Int. J. Hyg. Environ. Health, 207 (5), 409–417 (2004).
[34] *D.B. Barr, M.J. Silva, K. Kato, J.A. Reidy, N.A. Malek, D. Hurtz, M. Sadowski, L.L. Needham* and *A.M. Calafat:* Assessing Human Exposure to Phthalates Using Monoesters and Their Oxidized Metabolites as Biomarkers. Environ. Health Perspect. 111, 1148–1151 (2003).
[35] *H.A. Dirven, P.H. van den Broek* and *F.J. Jongeneelen:* Determination of four metabolites of the plasticizer di(2-ethylhexyl)phthalate in human urine samples. Int. Arch. Occup. Environ. Health. 64(8), 555–560 (1993).
[36] *Bundesärztekammer:* Qualitätssicherung der quantitativen Bestimmungen im Laboratorium. Neue Richtlinien der Bundesärztekammer. Dt. Ärztebl. 85, A699 – A712 (1988).
[37] *Bundesärztekammer:* Ergänzung der „Richtlinien der Bundesärztekammer zur Qualitätssicherung in medizinischen Laboratorien". Dt. Ärztebl. 91, C159 – C161 (1994).
[38] *G. Lehnert, J. Angerer* and *K.H. Schaller:* Statusbericht über die externe Qualitätssicherung arbeits- und umweltmedizinisch-toxikologischer Analysen in biologischen Materialien. Arbeitsmed. Sozialmed. Umweltmed. 33(1), 21–26 (1998).

[39] *J. Angerer* and *G. Lehnert.* Anforderungen an arbeitsmedizinisch-toxikologische Analysen – Stand der Technik. Dt. Ärztebl. 37, C1753 – C1760 (1997).

[40] *H.-D. Gilsing, J. Angerer* and *D. Prescher:* Convenient and high-yielding preparation procedures for mono(5-carboxy-2-ethylpentyl) phthalate and its ring-deuterated isomer – the "third" major metabolite of bis(2-ethylhexyl) phthalate. Monatsh. Chem. 136, 795–801 (2005).

Authors: *H. M. Koch, J. Angerer*
Examiners: *N. Bittner, E. R. Trösken, W. Völkel*

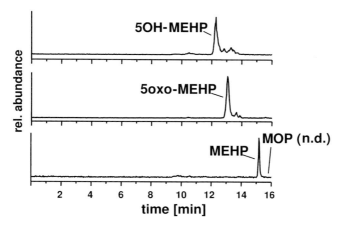

Fig. 4. Chromatogram of a worked-up urine sample. Only the quantifier traces of the parent-daughter ion combination of the analytes are shown. The concentrations were as follows: 5OH-MEHP: 41.4 µg/L; 5oxo-MEHP: 48.6 µg/L; MEHP: 15.0 µg/L; MOP: not detectable. The creatinine content was 0.72 g/L

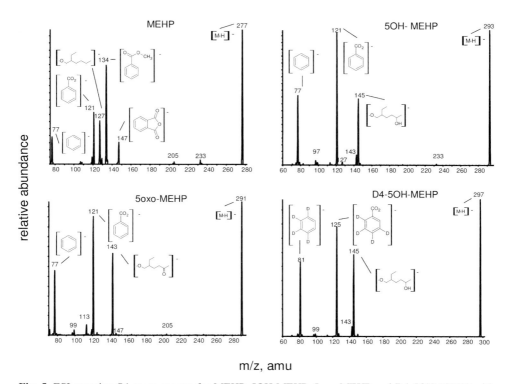

Fig. 5. ESI-negative Q1 mass spectra for MEHP, 5OH-MEHP, 5oxo-MEHP and D4-5OH-MEHP with the corresponding fragment structures

5-Hydroxy-N-methyl-2-pyrrolidone (5-HNMP) and 2-hydroxy-N-methylsuccinimide (2-HMSI)

Application   Determination in urine

Analytical principle   Capillary gas chromatography/ mass spectrometric detection (MS)

Summary

Both main metabolites of the working material N-methyl-2-pyrrolidone (NMP), i.e. 5-hydroxy-N-methyl-2-pyrrolidone (5-HNMP) and 2-hydroxy-N-methylsuccinimide (2-HMSI), are detected in urine by means of this GC/MS method.
5-HNMP and 2-HMSI labelled with deuterium are added to the urine samples as internal standards. Then the analytes are separated from interfering matrix components by means of solid phase extraction. The enriched analytes are eluted from the SPE columns using a mixture of ethyl acetate and methanol. The eluate is evaporated to dryness in a stream of nitrogen. Then bis-(trimethylsilyl)trifluoroacetamide is used to generate derivatives of the analytes. After capillary gas chromatographic separation, the analytes are determined by means of a mass selective detector and electron impact ionisation in the SIM mode. The concentration of the metabolites is calculated from calibration curves. The calibration curves are plotted by analysis of calibration standard solutions that are prepared in pooled urine and are treated in the same manner as the samples to be analysed.

5-Hydroxy-N-methyl-2-pyrrolidone (5-HNMP)

Within-series imprecision: Standard deviation (rel.) $s_w = 1.2\%$ or 0.9%
Prognostic range $u = 2.7\%$ or 2.0%
at a concentration of 7.5 mg or 75 mg 5-HNMP per litre urine and where n = 10 determinations

Between-day imprecision:	Standard deviation (rel.)	$s_w = 1.3\%$ or 2.3%
	Prognostic range	$u = 2.9\%$ or 5.1%
	at a concentration of 7.5 mg or 75 mg 5-HNMP per litre urine and where $n = 10$ determinations	
Accuracy:	Recovery rate	$r = 97\%$ at 150 mg/L
Detection limit:	1 mg 5-HNMP per litre urine	

2-Hydroxy-N-methylsuccinimide (2-HMSI)

Within-series imprecision:	Standard deviation (rel.)	$s_w = 1.4\%$ or 1.7%
	Prognostic range	$u = 3.1\%$ or 3.8%
	at a concentration of 7.5 mg or 75 mg 2-HMSI per litre urine and where $n = 10$ determinations	
Between-day imprecision:	Standard deviation (rel.)	$s_w = 3.7\%$ or 4.1%
	Prognostic range	$u = 8.3\%$ or 9.1%
	at a concentration of 5 or 75 mg 2-HMSI per litre urine and where $n = 10$ determinations	
Accuracy:	Recovery rate	$r = 101\%$ at 75 mg/L
Detection limit:	1 mg 2-HMSI per litre urine	

N-Methyl-2-pyrrolidone (NMP)

N-Methyl-2-pyrrolidone (NMP) (CAS No. 872-50-4, molecular weight 99.13 g/mol, melting point –24.4 °C, boiling point 204.3 °C, density 1.028 g/cm^3) is a heterocyclic ketone with properties that make it very well suited for use as a solubiliser, especially in aqueous/organic mixed phases. In particular, NMP is used as a solubiliser in agricultural chemicals or for cleaning purposes in the microelectronic industry and for paint stripping [1]. Approximately 38 000 tonnes of NMP are produced in the EU every year, about half that amount in Germany [2].

The MAK Commission have assigned a limit value of 20 mL/m^3 (equivalent to 82 mg/m^3) [3, 4]. A comprehensive review of the toxicological aspects of NMP can be found in the justification for the MAK value [4].

As the vapour pressure is relatively low under normal conditions (2.6 hPa), workplace concentrations at MAK value levels are reached only under unfavourable conditions, e.g. in the case of high surface temperatures and poor ventilation. However, due to ready dermal absorption of NMP, considerable exposure may occur that is not detected by measuring the air concentrations at the workplace, e.g. as a result of manual cleaning activities [5, 6].

NMP is readily absorbed by inhalation as well as through the skin, and it is rapidly distributed in the organism. After inhalation the retention of NMP is approx. 90%

[7]. According to the current state of scientific knowledge NMP is sequentially metabolised by oxidation in the liver [8, 9]. In addition to the main metabolites 5-hydroxy-*N*-methyl-2-pyrrolidone (5-HNMP, which amounts to approx. 68% of the metabolites excreted in urine) and 2-hydroxy-*N*-methylsuccinimide (2-HMSI, approx. 30% of the metabolites in urine) only low concentrations of *N*-methylsuccinimide (MSI) and unchanged NMP were found in the urine and in plasma (see Figure 1). Following inhalation the biological half-lives of 5-HNMP and 2-HMSI in urine are 7.3 h and 17 h respectively.

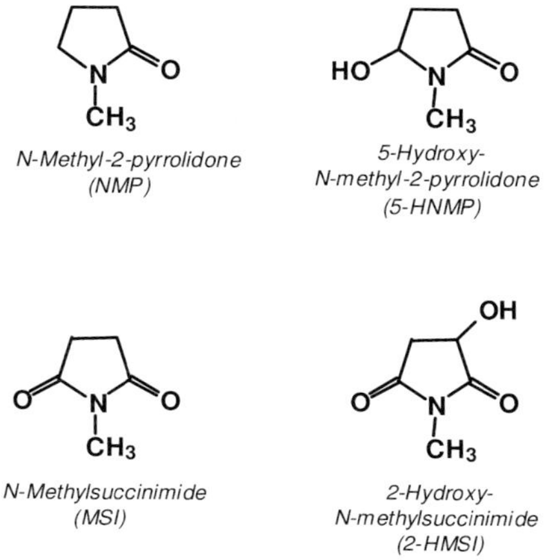

N-Methyl-2-pyrrolidone
(NMP)

5-Hydroxy-
N-methyl-2-pyrrolidone
(5-HNMP)

N-Methylsuccinimide
(MSI)

2-Hydroxy-
N-methylsuccinimide
(2-HMSI)

Fig. 1. Structural formulae of NMP and its main metabolites

Few experimental studies have been performed to investigate the relationship between the external levels of NMP and resulting internal exposure to the substance. According to Åkesson and Jönsson (2000) the 5-HNMP concentration in urine after exposure to NMP at the level of the MAK value for 8 h is approx. 200 mg/L and approx. 75 mg/L in the case of 2-HMSI [7]. The studies performed by Åkesson et al. show that there is a close correlation under experimental conditions between the external exposure level and the metabolites excreted in urine.

Anundi et al. and Langworth et al. carried out investigations on the exposure of personnel during the removal of graffiti from underground train stations. As the proportion of NMP in the solvents for removing paints and lacquers was only of minor importance (approx. 10% of the mass), it proved impossible to establish a quantitatively measurable relationship between the external and internal exposure [5, 6, 10]. When pure NMP was used to clean mixing vessels and tools, Bader et al. recorded average air exposure levels of 3 mg/m^3, and peaks up to 19 mg/m^3 were measured for short periods [11]. The mean excretion of 5-HNMP in the urine of the 10 investigated subjects was approx. 10 mg/L. In two cases skin contact was probable while carrying out intensive cleaning activities under unfavourable working conditions. Elevated internal exposure levels of 34 and 124 mg/L respectively were determined for the two workers.

Systematic investigations on the background exposure to N-methyl-2-pyrrolidone have not yet been carried out.

5-HNMP and 2-HMSI in urine seem to be the most suitable parameters to measure internal exposure levels to NMP, as determination of these metabolites cannot be affected by exogenic contamination.

Authors: *M. Bader, W. Rosenberger, W. Will*
Examiners: *J. Angerer, G. Müller, G. Leng*

5-Hydroxy-*N*-methyl-2-pyrrolidone (5-HNMP) and 2-hydroxy-*N*-methylsuccinimide (2-HMSI)

Application   Determination in urine

Analytical principle   Capillary gas chromatography/
                           mass spectrometric detection (MS)

Contents

1   General principles
2   Equipment, chemicals and solutions
2.1 Equipment
2.2 Chemicals
2.3 Solutions
2.4 Calibration standards
2.5 Conditioning of the solid phase extraction columns
3   Specimen collection and sample preparation
3.1 Sample preparation
4   Operational parameters
4.1 Operational parameters for gas chromatography and mass spectrometry
5   Analytical determination
6   Calibration
7   Calculation of the analytical result
8   Standardisation and quality control
9   Evaluation of the method
9.1 Precision
9.2 Accuracy
9.3 Detection limits
9.4 Sources of error
10 Discussion of the method
11 References

The MAK-Collection Part IV: Biomonitoring Methods, Vol. 11.
DFG, Deutsche Forschungsgemeinschaft
Copyright © 2008 WILEY-VCH Verlag GmbH & Co. KGaA, Weinheim
ISBN: 978-3-527-31596-3

Metabolites of *N*-methyl-2-pyrrolidone

1 General principles

5-HNMP and 2-HMSI labelled with deuterium are added to the urine samples as internal standards. Then the analytes are separated from interfering matrix components by means of solid phase extraction. The enriched analytes are eluted from the SPE columns using a mixture of ethyl acetate and methanol. The eluate is evaporated to dryness in a stream of nitrogen. Then bis-(trimethylsilyl)trifluoroacetamide is used to generate derivatives of the analytes. After capillary gas chromatographic separation, the analytes are determined by means of a mass selective detector and electron impact ionisation in the SIM mode. The concentration of the metabolites is calculated from calibration curves. The calibration curves are plotted by analysis of calibration standard solutions that are prepared in pooled urine and are treated in the same manner as the samples to be analysed.

2 Equipment, chemicals and solutions

2.1 Equipment

Gas chromatograph with split/splitless injector, mass selective detector (MSD), autosampler and data processing system

Capillary gas chromatographic column:
Length: 30 m; inner diameter: 0.25 mm; stationary phase: (50% phenyl)methylpolysiloxane; film thickness: 0.25 µm (e.g. J&W HP-50+, No. 19091L-433)

Solid phase extraction columns (SPE columns) ENV+ (200 mg, 6 mL) (e.g. Separtis, No. 915-0020-C)

Solid phase extraction unit (SPE station) with sealing valves (e.g. Varian VacElut)

Laboratory centrifuge (e.g. Heraeus Megafuge 1.0)

Solvent evaporator (e.g. from Pierce)

Drying cupboard (e.g. from Kendro-Heraeus)

Vortex mixer (e.g. from Ikamag)

6 mL Crimp-capped vials with crimp caps (e.g. from Macherey-Nagel)

10 mL Polyethylene centrifuge tubes (e.g. from Omnilab)

10 mL Test-tubes glass (e.g. from Omnilab)

1.8 mL Crimp-capped vials for the autosampler with crimp caps (e.g. from Macherey-Nagel)

Piston pipettes, adjustable between 50 µL and 5 mL (e.g. from Eppendorf-Netheler-Hinz)

10 mL, 50 mL and 100 mL Volumetric flasks

500 mL Measuring cylinder

Analytical balance (e.g. from Mettler-Toledo)

500 mL Glass bottles (e.g. from Omnilab)

13 mL Polypropylene tubes (e.g. from Sarstedt)

0.5 L and 2 L polyethylene bottles for collecting urine (e.g. from Sarstedt)

2.2 Chemicals

If not otherwise specified, all the chemicals must be at least p.a. grade.

Ethyl acetate (e.g. Merck No. 1.09623.1000)

Methanol, HPLC grade (e.g. J.T. Baker, No. 9093-02)

Water (ASTM type I) (e.g. Millipore Quantum purified)

5-Hydroxy-*N*-methyl-2-pyrrolidone (e.g. from Synthelec)

2-Hydroxy-*N*-methylsuccinimide (e.g. from Synthelec)

d_4-5-Hydroxy-*N*-methyl-2-pyrrolidone (e.g. from Synthelec)

d_3-2-Hydroxy-*N*-methylsuccinimide (e.g. from Synthelec)

N,O-bis(Trimethylsilyl)trifluoroacetamide (BSTFA) (e.g. Sigma-Aldrich, No. T 5634)

Nitrogen 5.0 (e.g. from Linde)

Helium 5.0 (e.g. from Linde)

2.3 Solutions

Elution solution:
400 mL ethyl acetate and 100 mL methanol are each measured in a measuring cylinder and mixed together in a 500 mL glass bottle.

2.4 Calibration standards

5-HNMP/2-HMSI stock solution:
Approx. 50 mg 5-HNMP and 25 mg 2-HMSI are weighed exactly in a 10 mL volumetric flask. The flask is subsequently filled to the nominal volume with ultra-pure water and the contents are mixed thoroughly (c = 5000 mg/L 5-HNMP and 2500 mg/L 2-HMSI).

The stock solution is stored in the deep-freezer at a maximum temperature of –18 °C and is stable for at least 1 year under these conditions.

The calibration standard solutions are prepared in pooled urine from non-exposed test persons. Various spontaneous urine samples are thoroughly mixed in a 2 L urine collection bottle to prepare the pooled urine.
As stipulated in the following pipetting scheme (see Table 1), calibration standard solutions are prepared in 100 mL volumetric flasks using the 5-HNMP/2-HMSI stock solution.

Table 1. Pipetting scheme for the preparation of the calibration standard solutions

Calibration standard solution No.	Volume of the stock solution [mL]	Volume of pooled urine [mL]	Concentration of 5-HNMP [mg/L]	Concentration of 2-HMSI [mg/L]
1	0	100	0	0
2	0.05	99.95	2.5	1.25
3	0.1	99.9	5	2.5
4	0.2	99.8	10	5.0
5	0.5	99.5	25	12.5
6	1	99	50	25
7	2	98	100	50
8	4	96	200	100

The calibration standard solutions thus prepared are divided into aliquots in 6 mL crimp-capped vials, sealed and stored in the deep-freezer at –18 °C until processing. They can be stored for at least 6 months under these conditions.

Solution of the internal standard (ISTD)

25 mg of d_4-5-hydroxy-N-methyl-2-pyrrolidone and d_3-2-hydroxy-N-methyl-succinimide are each weighed exactly in a 50 mL volumetric flask. The flask is subsequently filled to its nominal volume with ultrapure water and the contents are mixed thoroughly (c = 500 mg/L in each case).
The solution of the internal standard thus prepared is divided into aliquots in 6 mL crimp-capped vials, sealed and stored in the deep-freezer at –18 °C until processing. It can be stored for at least 6 months under these conditions.

2.5 Conditioning of the solid phase extraction columns

All the surfaces of the SPE station that come into contact with the sample and the eluent are cleaned with methanol before use.

The SPE columns are mounted on the SPE station and conditioned first with 5 mL methanol, then with 10 mL ultrapure water. The columns may not run dry during conditioning (use the sealing valves).

3 Specimen collection and sample preparation

The urine samples are collected in sealable 500 mL collection bottles and stored in the deep-freezer at a maximum temperature of $-18\,°C$ until processing. The urine can be stored in this manner for at least six months.

3.1 Sample preparation

Before analysis, the urine samples are thawed in a water bath at approx. $20\,°C$. As soon as they have reached room temperature they are thoroughly mixed.
1 mL of the urine sample is pipetted into a 10 mL glass tube. After addition of 100 µL of the solution of the internal standard, the sample is filled to a volume of 4 mL with ultrapure water and the contents of the tube are thoroughly mixed. Then 1 mL of this sample dilution is introduced onto an SPE column that has been preconditioned as described in Section 2.5. It is then sucked into the column by applying a slight vacuum on the SPE station. Then the column is washed with 1 mL water and dried for a period of 10 min by applying as high a vacuum as possible. Then the column is placed in a 10 mL centrifuge tube. It must be ensured that the centrifuge tube has a diameter that permits the column to be positioned in such a manner that its tip does not touch the bottom of the tube. The column is then centrifuged for 10 min at 3500 g. The valves, connections and delivery needles of the SPE station are cleaned with methanol.
A 10 mL glass tube is placed in the SPE station to collect the eluate. The SPE columns are mounted in the SPE station again for elution of the 5-HNMP and 2-HMSI metabolites and 2 mL of the elution mixture are added while the valve is open. The elution speed should be as slow as possible (approx. 1 drop per second) and therefore only a very slight vacuum, if any, should be applied. Then 1 mL of the eluate is transferred to a 1.8 mL sample vial and evaporated to dryness in a stream of nitrogen at a maximum temperature of $40\,°C$. 50 µL BSTFA are added to the residue, the vessel is tightly sealed and heated for 60 min at $110\,°C$ in the drying cupboard. After cooling to room temperature, 1 mL ethyl acetate is added. The sample vials are sealed, vigorously shaken, and 1 µL of the solution are analysed by means of GC/MS.

4 Operational parameters

4.1 Operational parameters for gas chromatography and mass spectrometry

Capillary column:	Material:	Fused silica
	Stationary phase:	HP 50+
	Length:	30 m
	Inner diameter:	0.25 mm
	Film thickness:	0.25 µm
Detector:	Mass selective detector (MSD)	
Temperatures:	Column:	Initial temperature 70 °C, then increase at a rate of 5 °C/min to 125 °C, then increase at a rate of 25 °C/min to 250 °C, then 5 minutes at the final temperature
	Injector:	230 °C
	Transfer line:	260 °C
Carrier gas:	Helium 5.0 at a constant flow rate of 1.2 mL/min	
Split:	1:20	
Sample volume:	1 µL	
Ionisation type:	Electron impact ionisation (EI)	
Ionisation energy:	70 eV	
Dwell time:	100 ms	
Electron multiplier:	Approx. 1400 V (autotuning value)	

All other parameters must be optimised in accordance with the manufacturer's instructions.

5 Analytical determination

The urine samples processed as described in Section 3.1 are analysed by injecting 1 µL of the ethyl acetate extract into the gas chromatograph. The temporal profiles of the ion traces shown in Table 2 are recorded in the SIM mode (see also Figures 2, 3 and 4):

The retention times shown in Table 2 serve only as a guide. Users of the method must satisfy themselves of the separation power of the capillary column used and the resulting retention behaviour of the substances.

Table 2. Retention times and recorded masses

Compound	Retention time [min]	Mass [m/z]
5-HNMP	12.53	186* 172
d_4-5-HNMP	12.48	190* 176
2-HMSI	12.60	186* 144
d_3-2-HMSI	12.56	189* 147

The masses marked * are used for quantitative evaluation.

If the measured values are above the linear range of the calibration graphs (i.e. >200 mg/L), then the urine samples are appropriately diluted with ultrapure water, processed anew and analysed. Two quality controls are included in each series.

6 Calibration

The calibration standard solutions prepared as described in Section 2.4 are processed in the same manner as the urine samples (Section 3.1) and analysed as stipulated in Sections 4 and 5. The calibration graph is obtained by plotting the quotients of the peak areas of the 5-HNMP derivative or the 2-HMSI derivative with the appropriate internal standard as a function of the concentrations used (see Figure 5). It is not necessary to plot a complete calibration graph for every analytical series. It is sufficient to analyse one calibration standard for every analytical series. The ratio of the result obtained for this standard and the result for the equivalent standard in the complete calibration graph is calculated. Each result that is read off the calibration graph is corrected using this quotient (one-point calibration).
New calibration graphs should be plotted if the quality control results indicate systematic deviation.
The calibration curve is linear between the detection limit (1 mg/L) and 200 mg/L.

7 Calculation of the analytical result

The concentrations of 5-HNMP and 2-HMSI in the sample are calculated using the quotients of the peak areas of the analyte and the internal standard obtained from the sample to be investigated and from the calibration function. If dilution has been previously carried out, this must be taken into account by multiplying the analytical result by the dilution factor.

8 Standardisation and quality control

Quality control of the analytical results is carried out as stipulated in the guidelines of the Bundesärztekammer (German Medical Association) [12] and in the special preliminary remarks to this series. Two spiked urine samples are included to check the precision. The control material can be prepared in the laboratory or obtained from commercial suppliers (e.g. Recipe, Munich, Germany). Control samples are prepared using pooled urine spiked with defined amounts of 5-HNMP and 2-HMSI in the decisive concentration range (approx. 25 mg/L for low NMP exposure levels to approx. 175 mg/L for high NMP exposure levels). A six-month supply of the control material is prepared, divided into aliquots in 6 mL crimp-capped vials and stored in the deep-freezer. The control material is prepared freshly every six months. The expected values and the tolerance range of this quality control material are determined in a pre-analytical period (one analysis on each of 20 different days) [13–15].

9 Evaluation of the method

9.1 Precision

Pooled urine from non-exposed test persons is used to determine the precision in the series. This was spiked with the quantities of 5-HNMP and 2-HMSI shown in Table 3, and then processed and analysed as described in the previous Sections in 10 consecutive measurements. The precision from day to day was checked by analysing these samples on 10 different days (see Table 3).

Table 3. Precision for the determination of 5-HNMP and 2-HMSI

	n	Spiked concentration [mg/L]	Standard deviation (rel.) [%]	Prognostic range [%]
5-HNMP				
In the series	10	7.5	1.2	2.7
		75	0.9	2.0
From day to day	10	7.5	1.3	2.9
		75	2.3	5.1
2-HMSI				
In the series	10	7.5	1.4	3.1
		75	1.7	3.8
From day to day	10	5	3.7	8.2
		75	4.1	9.1

9.2 Accuracy

The accuracy of the method was checked by recovery experiments and by participation in external quality assurance programmes (round-robin experiments organised by the *Deutsche Gesellschaft für Arbeitsmedizin und Umweltmedizin*/German Society for Occupational Medicine and Environmental Medicine (c/o Institute of Occupational, Social and Environmental Medicine of the University of Erlangen-Nürnberg)).
Ten different spiked urine samples (150 mg/L 5-HNMP, 75 mg/L 2-HMSI) were processed in one analytical series. Mean values of 146±12 mg/L (5-HNMP) and 76±3 mg/L (2-HMSI) were found. These results lead to relative recovery rates of 97±8% (5-HNMP) and 101±4% (2-HMSI).
A mean recovery of 88% relative to the expected result was determined for 5-HNMP concentrations of between 51 and 100 mg/L in two round-robin tests. In the case of 2-HMSI the mean recovery was 100% in a range between 31 and 50 mg/L.

9.3 Detection limits

The detection limit of the method in urine is 1 mg/L for both substances, the quantitation limit is 2 mg/L. Both were determined with the calibration point procedure according to DIN 32 645 [16].

9.4 Sources of error

The determination of 5-HNMP and 2-HMSI may be considerably influenced by three factors:
- The consistency of the urine matrix has a strong influence on the absolute recovery from solid phase extraction. The losses due to processing increase as the creatinine content of the urine rises (>1.5 g/L), which also results in diminished precision.
- A precondition for derivatisation using BSTFA is the absence of water. Therefore the solid phase columns must be dried as completely as possible, as otherwise the subsequent derivatisation with BSTFA will be incomplete.
- As a rule, measurement signals that show no differences in their retention behaviour or their fragmentation patterns from those of 5-HNMP and 2-HMSI are observed in all urine samples below the arithmetically calculated detection limits. This is possibly caused by contamination of the deuterated internal standards by the non-deuterated metabolites.

10 Discussion of the method

This method ensures detection of occupational exposure to NMP in the linear range between the detection limit of 1 mg/L and at least 200 mg/L. The use of isotope-labelled internal standards leads to reliable analytical determination of the analytes. Moreover, fluctuations due to the matrix, e.g. the influence of different urine matrices, or due to processing are compensated, so that very good precision is achieved by the method. Both the values for the precision in the series and from day to day with a maximum of 4.1% are considered very good, and the examiners achieved similar results.

The method described here is based on a procedure developed by Jönsson and Åkesson for simultaneous determination of 5-HNMP and 2-HMSI in urine [17]. On account of the high metabolite concentrations that occur following relatively low-level NMP exposure several dilution steps were introduced compared with the original procedure (dilution of the urine, use of only part of the solid phase eluate, reconstitution of the sample in 1 mL ethyl acetate, split injection) to shift the mass spectrometric signal into the linear range. These changes resulted in a detection limit that was about 10 times higher than that given by Jönsson and Åkesson (0.1 mg/L) [17]. The analytical advantages of the higher dilution lie in the optimisation of the method precision and minimisation of the matrix-dependent effects. Furthermore, higher sensitivity is not required in practice in occupational medicine, as NMP exposure below a level of 5% of the MAK value results in two-digit metabolite concentrations in the mg/L range [11]. If necessary, the detection sensitivity can be increased by omitting one or several of the dilution steps.

As an alternative to GC/MS Carnerup et al. have established an HPLC/MS/MS procedure that exhibits comparable precision data and detection limits [18]. The main advantage of this method is the simplified sample preparation, as no derivatisation of the metabolites is performed.

C_8-endcapped material was used for solid phase extraction in the original method by Jönsson and Åkesson [17]. Carnerup et al. modified the extraction by using cross-linked polystyrene-divinylbenzene resins and achieved recovery rates of almost 100% for both analytes after elution with ethyl acetate/ethanol [18]. Moreover, during validation of this method it was observed that, compared with the reversed-phase cartridges, more complete and more reliable drying of the mixed resin cartridge was achieved by vacuum and centrifugation. The resin matrix can also be dried by sucking air through the cartridge for 30 min on the SPE station as an alternative to centrifugation.

From the occupational medicine point of view simultaneous determination of both main metabolites of NMP has the advantage of enabling investigation of two biomarkers with different elimination behaviour. Whereas the excretion of 5-HNMP reaches its highest value in the first two hours after exposure, 2-HMSI has an elimination maximum of 16 to 20 hours after the end of exposure and a half-life of approx. 17 hours and exhibits accumulation of the working material for several days. It is to be expected that any individual influences (e.g. drinking behaviour, temporary exposure peaks) will tend to be levelled out when 2-HMSI is measured. This permits a more representative assessment of exposure on account of the longer elimination of the metabolite. However, as the proportion of 5-HNMP in the total excretion of the NMP metabolites in urine is about 2/3, this parameter is possibly more diagnostically sensitive than 2-HMSI.

Instruments used:
Gas chromatograph 6890 with mass selective detector 5973, autosampler and data system from Agilent

11 References

[1] *Health and Safety Executive (HSE):* N-Methyl-2-pyrrolidone. Risk assessment document. EH72/10, ISBN 0717615286, HSE Books, Sudbury, UK (1997).
[2] *M. Bader, W. Will, R. Rossbacher, M. Nasterlack* and *R. Wrbitzky:* N-Methyl-2-pyrrolidon – Arbeitsmedizinische Bedeutung und Toxikologie. Arbeitsmed. Sozialmed. Umweltmed. 9, 422–428 (2002).
[3] *Deutsche Forschungsgemeinschaft:* List of MAK and BAT Values 2006, 42nd report, Wiley-VCH, Weinheim (2006).
[4] *H. Greim (ed.):* N-Methyl-2-pyrrolidone. Occupational Toxicants – Critical Data Evaluation for MAK Values and Classification of Carcinogens. Issues 20, 34, 41. Wiley-VCH, Weinheim (1994, 2002, 2006).
[5] *H. Anundi, M.L. Lind, L. Friis, N. Itkes, S. Langworth* and *C. Edling:* High exposures to organic solvents among graffiti removers. Int. Arch. Occup. Environ. Health 65, 247–251(1993).
[6] *S. Langworth, H. Anundi, L. Friis, G. Johanson, M.L. Lind, E. Söderman* and *B. Åkesson:* Acute health effects common during graffiti removal. Int. Arch. Occup. Environ. Health 74, 213–218 (2001).
[7] *B. Åkesson* and *B.A.G. Jönsson:* Biological monitoring of N-methyl-2-pyrrolidone using 5-hydroxy-N-methyl-2-pyrrolidone in plasma and urine as the biomarker. Scand. J. Work Environ. Health 26, 213–218 (2000).
[8] *D.A. Wells, A.A. Hawi* and *G.A. Digenis:* Isolation and identification of the major urinary metabolite of N-methylpyrrolidone in the rat. Drug Metab. Dispos. 20, 124–126 (1992).
[9] *B. Åkesson* and *B.A. Jönsson:* Major metabolic pathway for N-methyl-2-pyrrolidone in humans. Drug Metab. Dispos. 25, 267–269 (1997).
[10] *H. Anundi, S. Langworth, G. Johanson, M.L. Lind, B. Åkesson, L. Friis, N. Itkes, E. Söderman, B.A.G. Jönsson* and *C. Edling:* Air and biological monitoring of solvent exposure during graffiti removal. Int. Arch. Occup. Environ. Health 73, 561–569 (2000).
[11] *M. Bader, W. Rosenberger, Th. Rebe, S. Keener, T.H. Brock, H. Hemmerling* and *R. Wrbitzky:* Ambient monitoring and biomonitoring of workers exposed to N-methyl-2-pyrrolidone (NMP) in an industrial facility. Int. Arch. Occup. Environ. Health 79, 357–364 (2005).
[12] *Bundesärztekammer:* Richtlinie der Bundesärztekammer zur Qualitätssicherung quantitativer laboratoriumsmedizinischer Untersuchungen. Dt. Ärztebl. 100, A3335 – A3338 (2003).
[13] *J. Angerer* and *G. Lehnert:* Anforderungen an arbeitsmedizinisch-toxikologische Analysen – Stand der Technik. Dt. Ärztebl. 37, C1753–C1760 (1997).
[14] *J. Angerer, Th. Göen* and *G. Lehnert:* Mindestanforderungen an die Qualität von umweltmedizinisch-toxikologischen Analysen. Umweltmed. Forsch. Prax. 3, 307–312 (1998).
[15] *G. Lehnert, J. Angerer* and *K.H. Schaller:* Statusbericht über die externe Qualitätssicherung arbeits- und umweltmedizinisch-toxikologischer Analysen in biologischen Materialien. Arbeitsmed. Sozialmed. Umweltmed. 33(1), 21–26 (1998).
[16] *DIN 32 645:* Nachweis-, Erfassungs- und Bestimmungsgrenze. Beuth Verlag, Berlin (1994).
[17] *B.A.G. Jönsson* and *B. Åkesson:* Determination of 5-hydroxy-N-methylpyrrolidone and 2-hydroxy-N-methylsuccinimide in human urine. J. Chrom. B 694, 351–357 (1997).
[18] *M.A. Carnerup, B. Åkesson* and *B.A.G. Jönsson:* Determination of 5-hydroxy-N-methyl-2-pyrrolidone and 2-hydroxy-N-methylsuccinimide in human plasma and urine using liquid chromatography-electrospray tandem mass spectrometry. J. Chrom. B 761, 107–113 (2001).

Authors: *M. Bader, W. Rosenberger, W. Will*
Examiners: *J. Angerer, G. Müller, G. Leng*

Metabolites of N-methyl-2-pyrrolidone

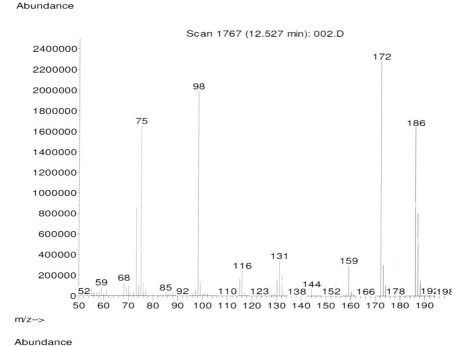

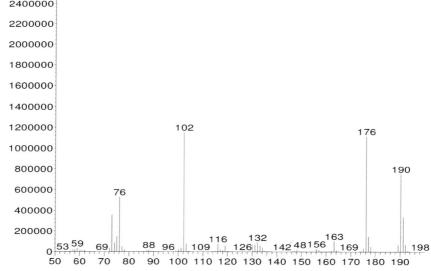

Fig. 2. Mass spectra of 5-HNMP (top) and d_4-5-HNMP (bottom)

Metabolites of N-methyl-2-pyrrolidone

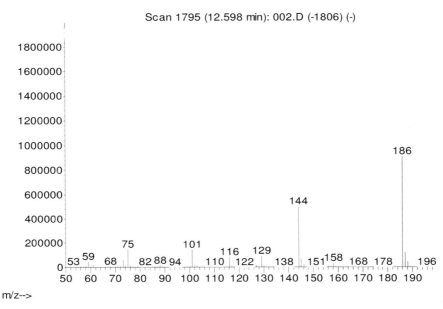

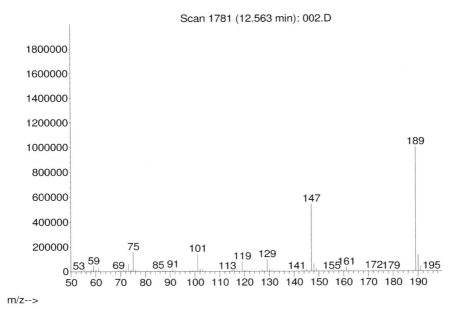

Fig. 3. Mass spectra of 2-HMSI (top) and d_3-2-HMSI (bottom)

Metabolites of *N*-methyl-2-pyrrolidone

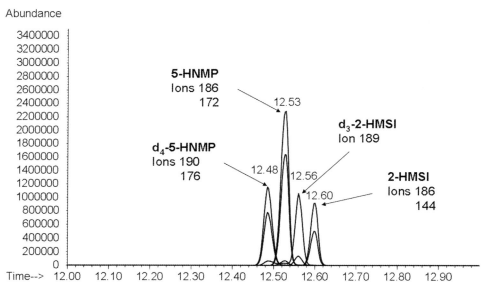

Fig. 4. Gas chromatogram with ion traces of the NMP metabolites and the internal standard (c = 200 mg/L 5-HNMP (t_R = 12.53 min) (d_4-5-HNMP = 12.48 min), 100 mg/L 2-HMSI (t_R = 12.60 min) (d_3-2-HMSI = 12.56 min)

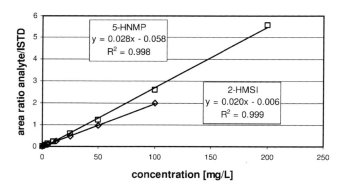

Fig. 5. Calibration functions for 5-HNMP and 2-HMSI

Iridium

Application   Determination in urine

Analytical principle   Sector field mass spectrometry with inductively coupled plasma

Summary

The sector field mass spectrometric method with inductively coupled plasma (SF-ICP-MS) in the low-resolution mode that is described here is suitable for detecting the occupational and environmental exposure to iridium in urine. In principle, it is equivalent to the method for the determination of platinum and gold in urine.
In this case the urine samples are subjected to UV digestion and then introduced into the ICP-MS using a Meinhard nebulizer. Evaluation is carried out using the standard addition procedure.

Iridium

Within-series imprecision: Standard deviation (rel.) $s_w = 4.3\%$
 Prognostic range $u = 11.1\%$
 at a concentration of 1.1 ng per litre urine
 and where n = 5 determinations

Between-day imprecision: Standard deviation (rel.) $s_w = 7.3\%$
 Prognostic range $u = 18.8\%$
 at a concentration of 1.4 ng per litre urine
 and where n = 5 determinations

Accuracy: Recovery rate $r = 98\%$

Detection limit: 0.1 ng iridium per litre urine

Iridium

Iridium (Ir; atomic number 77; relative atomic mass: 192.217) belongs to the platinum group of elements. It is composed of two naturally occurring isotopes with atomic masses of 191 (relative frequency: 37.3%) and 193 (relative frequency: 62.7%). Moreover, several synthetic isotopes are known. They have atomic masses of between 182 and 198 and their half-lives range from 4.9 s to 74 days. Iridium can occur in all the oxidation states between 0 and +VI, whereby oxidation states 0, +III and +IV are the most common and the most stable.

As iridium is brittle, it is primarily used in alloys, which serve as starting materials for the manufacture of jewellery, injection needles, instrument parts, electrical contacts, dental alloys, etc. Furthermore, iridium serves as a catalyst in the chemical industry. ^{192}Ir is used for radiation therapy in cases of cancer.

Little information on the background exposure of the general population to iridium has been published in the literature, and no studies have been performed on the internal exposure of those who handle iridium at the workplace. Table 1 lists the available data on iridium excretion in urine. The most comprehensive investigation was carried out by the *Umweltbundesamt* [German Federal Environment Agency] as part of the Environmental Survey in 1998. A mean iridium excretion of 0.24 ng/L was found in the urine of 1080 adults aged between 18 and 69 who were not occupationally exposed to the element, the 95th percentile was 1.1 ng/L [1].

Table 1. Concentrations of iridium in the urine of various population groups

Group of persons	N	Mean value, range [ng/L]	References
General population in Germany, 18 to 69 years old	1080	0.24, <0.1–16.5	[1]
11 to 48 years old	19	<0.9–2.1	[2]
Unknown	17	0.7–70	[3]

No investigations on the effects of chronic exposure of humans and the environment to iridium are available to date.

Authors: *J. Begerow, L. Dunemann*
Examiner: *P. Schramel*

Iridium

Application　　　　Determination in urine

Analytical principle　Sector field mass spectrometry
　　　　　　　　　　　with inductively coupled plasma

Contents

1　General principles
2　Equipment, chemicals and solutions
2.1　Equipment
2.2　Chemicals
2.3　Solutions
2.4　Calibration standards
3　Specimen collection and sample preparation
3.1　Sample preparation
4　Operational parameters
5　Analytical determination
6　Calibration
7　Calculation of the analytical result
8　Standardisation and quality control
9　Evaluation of the method
9.1　Precision
9.2　Accuracy
9.3　Detection limits
9.4　Sources of error
10　Discussion of the method
11　References

1 General principles

The urine samples are subjected to UV digestion and then introduced into the ICP-MS using a Meinhard nebulizer. The evaluation is carried out using the standard addition procedure.

2 Equipment, chemicals and solutions

2.1 Equipment

Double-focussing sector field ICP-MS with autosampler and peristaltic pump

Air-conditioned room (18 to 22 °C) with particle filter (separation efficiency ≥99.997% for particles with a diameter of 0.3 µm)

Cleanroom workstation (clean bench, laminar flow)

UV digestion device with 25 mL quartz vessels and 1000 W medium-pressure mercury vapour lamp (e.g. UV 1000, Kürner)

Microlitre pipettes, adjustable between 10 and 100 µL, and between 100 and 1000 µL (e.g. from Eppendorf)

Millilitre pipette, adjustable between 1 and 10 mL (e.g. Varipette, Eppendorf)

250 mL Polypropylene vessels with screw caps (e.g. from Sarstedt)

15 mL Polypropylene centrifuge tubes with polypropylene stoppers (e.g. from Sarstedt)

10 mL, 100 mL and 1000 mL Volumetric flasks

Quartz apparatus for subboiling distillation (e.g. from Kürner)

500 mL Measuring cylinder

2.2 Chemicals

Iridium standard solution (1 g/L) in 10% HCl (e.g. Spex, No. PLIR3-2Y).

Ultrapure water (equivalent to ASTM type 1) or double-distilled water

Argon 5.0 (e.g. from Linde)

Concentrated nitric acid (e.g. Trace SELECT^{TM} ≥69.0%, Fluka No. 84385) additionally purified by subboiling distillation

30% Hydrogen peroxide "Suprapur" (e.g. Merck, No. 1.07298.1000)

2.3 Solutions

Nitric acid of the highest degree of purity that is commercially available is further purified by subboiling distillation in a quartz apparatus.

Nitric acid diluted 1 + 2 (to clean the plastic vessels):
600 mL ultrapure water are placed in a 1000 mL volumetric flask. Then 300 mL of 69.0% nitric acid are added.

0.5% Nitric acid:
7.2 mL of 69.0% nitric acid are pipetted into a 1000 mL volumetric flask into which approx. 500 mL ultrapure water have already been placed. The volumetric flask is subsequently filled to its nominal volume with ultrapure water.
This solution must be stored in the refrigerator and must be freshly prepared each day.

2.4 Calibration standards

Starting solution:
100 µL of the iridium standard solution are pipetted into a 1000 mL volumetric flask, into which approximately 500 mL of 0.5% nitric acid have already been placed. The volumetric flask is subsequently filled to its nominal volume with 0.5% nitric acid (0.1 mg/L).
The starting solution can be stored in the refrigerator at +4 to +8 °C for about 4 weeks.

Stock solution:
1 mL of the starting solution is pipetted into a 1000 mL volumetric flask, into which approximately 500 mL of 0.5% nitric acid have already been placed. The volumetric flask is subsequently filled to its nominal volume with 0.5% nitric acid (0.1 µg/L).
The stock solution must be freshly prepared every working day.

Calibration standard solutions:
Calibration standards containing 2 to 25 ng iridium/L are prepared from this stock solution by dilution with 0.5% HNO_3. Table 2 shows the pipetting procedure to be followed.

Table 2. Pipetting scheme for the preparation of the calibration standard solutions

Volume of the stock solution [mL]	Final volume of the calibration standard solution [mL]	Concentration of the calibration standard [ng/L]	Designation
2	100	2	Std. I
5	100	5	Std. II
25	100	25	Std. III

The calibration standards must also be freshly prepared on each working day.

3 Specimen collection and sample preparation

As is the case for all trace element analyses, it is essential to ensure that the reagents are of the highest possible purity and that the vessels are thoroughly clean.
Before using the plastic vessels for sample collection and sample preparation, they must be cleaned by placing them in a bath of 1 + 2 diluted nitric acid for 30 min at 50 °C to prevent any possible extraneous contamination. Then the vessels are rinsed three times with ultrapure water and dried in the drying cupboard at 50 °C.
The samples of morning urine are collected directly in the clean polyethylene vessels. If the determination cannot be carried out immediately, the urine can be stored in the refrigerator for 1 to 2 days at +4 to +8 °C, but the urine should be acidified (approx. 10 mL concentrated nitric acid per litre urine). The samples can be kept for at least 6 months in the deep-freezer at –18 °C to –22 °C.
If applicable, the urine samples are thawed overnight for further processing and, as soon as they have reached room temperature, they are worked up as follows.

3.1 Sample preparation

Ultrapure water is processed as a reagent blank instead of the urine samples. As in the case of the urine samples, the water is acidified with 10 mL concentrated nitric acid per litre.
Before an aliquot is withdrawn for UV digestion, the urine samples are thoroughly shaken to ensure that any sediment is homogeneously distributed. Then 5 mL of the urine are pipetted into the 25 mL quartz tube, and 200 µL hydrogen peroxide are added. The quartz tube is placed in the UV digestion device and subjected to UV light for 15 min with the cooling water flowing at a rate of 0.9 mL per minute. The addition of 200 µL hydrogen peroxide and the 15-minute UV digestion are repeated twice (total digestion time 45 min). Then the quartz tube is removed from the digestion device and cooled at room temperature. The colourless to slightly yellow digestion solution is subsequently transferred to a 10 mL volumetric flask. The volumetric flask is then filled to its nominal volume with 0.5% nitric acid and the contents are

Table 3. Pipetting scheme for the standard addition procedure

Digestion solution [mL]	Water [mL]	0.5% Nitric acid [mL]	Std. I (2 ng/L) [mL]	Std. II (5 ng/L) [mL]	Std. III (25 ng/L) [mL]	Added Iridium concentration* [ng/L]
0.5	1.0	0.5	–	–	–	–
0.5	1.0	–	0.5	–	–	4
0.5	1.0	–	–	0.5	–	10
0.5	1.0	–	–	–	0.5	50

* These concentration data are based on a urine volume of 0.25 mL, as the urine samples are diluted 1+1 after UV digestion.

shaken thoroughly. 0.5 mL of the diluted digestion solution is pipetted from the 10 mL volumetric flask into a 5 mL autosampler vial. The sample is diluted with 1 mL ultrapure water, and 0.5 mL of the appropriate calibration standard solution (see Table 2) is added. Table 3 shows the pipetting procedure to be followed. Four solutions are thus obtained for each sample and they are analysed as described in Sections 4 and 5.

4 Operational parameters

Principle:	Double-focussing sector field ICP-MS
Autosampler:	Gilson type 222 XL
Peristaltic pump:	Gilson type Maniples 3: 1.4 mL/min
Resolution:	$m/\Delta m = 300$
Frequency generator:	1.30 kW
Scan mode:	Electrical scan
Integration time:	50 ms
Torch:	Fassel torch
Nebulizer:	Meinhard TR-30-A3
Spray chamber:	Scott type ("double-pass") at 20 °C
Cones:	Ni ("High performance")
Auxiliary gas:	Argon, 0.8 L/min
Cooling gas:	Argon, 13.3 L/min
Nebulizer gas:	Argon, 0.7 L/min
Purge time between samples:	3 min
Mass ranges:	190.6–191.3 (^{191}Ir)
	192.6–193.3 (^{193}Ir)
Passes:	5
Runs:	2
Sample time:	0.025 s
Scan time:	8 s
Total number of samples:	480

The given ICP-MS conditions serve only as a guideline. The individual parameters must be optimally adjusted for the instrument the user has available.

5 Analytical determination

The digestion solutions prepared as described in Section 3.1 (four per sample, see Table 3) are analysed on the basis of the instrumental parameters given above and the peak areas are recorded. On principle, duplicate determination is performed. A reagent blank and a quality control sample are included in each analytical series.

6 Calibration

7 Calculation of the analytical result

The iridium content of the urine sample is determined graphically with the help of the standard addition procedure. The mean peak areas of the unspiked and the three spiked samples calculated by the computer program and corrected by the blank value are plotted versus the iridium concentrations. The result is a linear graph and its point of interception with the concentration axis gives the iridium content in ng per litre. As a rule, sector field ICP-MS instruments are equipped with computer-supported evaluation programs that can perform this evaluation automatically.
The working range is linear from the detection limit (0.1 ng/L) to 5000 ng/L.

8 Standardisation and quality control

Quality control of the analytical results is carried out as stipulated in the guidelines of the Bundesärztekammer (German Medical Association) [4] and in the special preliminary remarks to this series. As no reference material for iridium is commercially available, it must be prepared in the laboratory. For this purpose pooled urine is spiked with a defined amount of iridium in the middle of the typically occurring concentration range (e.g. 1 ng/L). This control material is divided into aliquots in polypropylene bottles or tubes and can be stored at $-18\,°C$ to $-22\,°C$ for 6 months. The expected value and the tolerance range of this quality control material are determined in a pre-analytical period (one analysis of the control material on each of 20 different days) [5–7].

9 Evaluation of the method

9.1 Precision

The urine sample of a person who was not exposed to iridium at the workplace was used to check the precision in the series. This urine sample contained iridium at a mean concentration level of 1.1 ng/L. The relative standard deviation of the results of the 5-fold determination was 4.3%, which is equivalent to a prognostic range of 11.1%.

A native urine sample with a mean iridium content of 1.4 ng/L was processed and analysed as described in the preceding sections on five consecutive days to determine the precision from day to day. The standard deviation of the results was 7.3%, which is equivalent to a prognostic range of 18.8%.

9.2 Accuracy

The accuracy was evaluated by recording the results for two isotopes (^{191}Ir and ^{193}Ir). Spectral interference can be largely eliminated when the deviation from the natural isotope ratio lies within the range of the measurement imprecision.

A urine sample of a person who was not occupationally exposed to iridium was spiked with 5 ng/L iridium, and the sample was processed and analysed five times. The mean relative recovery was 98% in this case.

9.3 Detection limits

Under the given analytical conditions the detection limit was 0.1 ng iridium per litre urine. The detection limit was calculated as three times the standard deviation of 10 replicate measurements of the reagent blank value.

9.4 Sources of error

When extremely low concentrations are determined, as in the case of this method, it is especially important to avoid contamination caused by the reagents, the air in the laboratory, the vessels and pipettes used, etc. For example, it is important to carry out all the working steps under cleanroom conditions. It is essential to keep a continuous and intensive check on the reagent blank value.

Memory effects can occur when precious metals are analysed by ICP-MS. Therefore it is important to ensure that carry-over contamination from one measured sample to another does not take place. Samples containing higher concentrations (e.g. those of relevance to occupational medicine) should not be analysed in the same series with samples containing low concentrations (e.g. those relevant to environmental medicine). In the case of unknown samples it is helpful to carry out a preliminary analysis after dilution in order to estimate the concentration range. This can prevent contamination within the ICP-MS instrument (torch, spray chamber, tubes, etc.) that would entail laborious cleaning activities.

Interferences can occur in the ICP-MS for various reasons. Essentially, interference is divided into spectral and non-spectral sources of interference [8].

Spectral overlap due to isobar, polyatomic and multi-charged ions are of no significance in the determination of iridium in urine, so that maximum detective power can be achieved for measurement of the two available isotopes ($^{191}\text{Ir}^+$, $^{193}\text{Ir}^+$) in the low-resolution mode. On principle, however, sample digestion to destroy the organic matrix is strongly recommended when human biological samples are to be determined by ICP-MS. This distinctly reduces spectral and non-spectral interference and considerably improves the long-term stability of the ICP-MS. Furthermore, dilution of the urine samples and the use of the standard addition method effectively diminish non-spectral interference.

It is highly recommended to shake the samples vigorously after they have been stored, as any sedimentation in the urine samples can lead to adsorption of the analyte on the surface of the sediment, so that erroneous measurements cannot be ruled out.

10 Discussion of the method

The analytical procedure described here is an extremely powerful and reliable method of determination of this noble metal in urine both in the range of relevance to occupational medicine as well as that of interest to environmental medicine.

If required, this method can also be used to simultaneously assay further elements, such as platinum and gold and/or thorium and uranium (see also the "Platinum and gold" and "Thorium and uranium" methods in volumes 7 and 6 of this series) as well as iridium. As complicated processing and evaporation steps can be dispensed with and as several analytes are detectable in one analytical step, the method described here has proved extremely practicable and meets the requirements of occupational and environmental medicine for monitoring internal exposure to iridium.

The preceding UV digestion ensures that organic matrix components are removed and that the matrix load of the samples is effectively lowered. The advantage of this digestion method in particular over other frequently used procedures, such as thermal or microwave-induced high-pressure digestion, is that addition of reagents is minimised [9, 10]. Thus blank values caused by reagents can be kept to an absolute minimum and dilution of the sample to lower the acid concentration becomes superfluous. On principle, the urine samples can be analysed directly after dilution without previous digestion, but the long-term stability of the ICP-MS is not assured when larger series of samples are measured. Direct measurement is always possible if individual samples or only a few samples are to be analysed.

The quadrupole ICP-MS technique may be used as an alternative and modern instruments are capable of determination at concentrations of approx. 2 ng/L, making the concentration range of interest to occupational medicine accessible as well as the upper concentration range of relevance to environmental medicine.

Instruments used:
ELEMENT sector field ICP-MS (Finnigan MAT, Germany) with CD-1 system and ASX-400 autosampler (Cetac, USA), PC and printer.

11 References

[1] *K. S. Becker, S. Kaus, C. Krause, P. Lepom, C. Schulz, M. Seiwert* and *B. Seifert:* Umwelt-Survey 1998, Vol. III: Human-Biomonitoring. Stoffgehalte in Blut und Urin der Bevölkerung in Deutschland. Umweltbundesamt (2002).

[2] *I. Rodushkin* and *F. Ödman:* Application of inductively coupled plasma sector field mass spectrometry for elemental analysis in urine. J. Trace elements Med. Biol. 14, 241–247 (2001).

[3] *C. Minoia, E. Sabbioni, P. Apostoli, R. Pietra, L. Pozzoli, M. Gallorini, G. Nicolaou, L. Alessio* and *E. Capodaglio:* Trace element reference values in tissues from inhabitants of the European community I: A study of 46 elements in urine, blood, and serum of Italian subjects. Sci. Total Environ. 95, 89–105 (1990).

[4] *Bundesärztekammer:* Richtlinie der Bundesärztekammer zur Qualitätssicherung quantitativer laboratoriumsmedizinischer Untersuchungen. Dt. Ärztebl. 100, A3335–A3338 (2003).

[5] *J. Angerer* and *G. Lehnert:* Anforderungen an arbeitsmedizinisch-toxikologische Analysen – Stand der Technik. Dt. Ärztebl. 37, C1753–C1760 (1997).

[6] *J. Angerer, Th. Göen* and *G. Lehnert:* Mindestanforderungen an die Qualität von umweltmedizinisch-toxikologischen Analysen. Umweltmed. Forsch. Prax. 3, 307–312 (1998).

[7] *G. Lehnert, J. Angerer* and *K. H. Schaller:* Statusbericht über die externe Qualitätssicherung arbeits- und umweltmedizinisch-toxikologischer Analysen in biologischen Materialien. Arbeitsmed. Sozialmed. Umweltmed. 33(1), 21–26 (1998).

[8] *P. Schramel, J. Begerow* and *H. Emons:* Die Anwendung der ICP-MS für das Humanbiomonitoring. In: *J. Angerer* and *K. H. Schaller* (eds.): Analysen in biologischem Material, 13th issue, Deutsche Forschungsgemeinschaft, Wiley-VCH, Weinheim (1999).

[9] *J. Begerow, M. Turfeld* and *L. Dunemann:* Determination of physiological platinum levels in human urine using double focusing magnetic sector field inductively coupled plasma mass spectrometry in combination with ultraviolet photolysis. J. Anal. At. Spectrom. 1, 913–916 (1997).

[10] *E. Saur* and *E. Spahn:* Die UV-Photolyse – ein nahezu reagenzienfreies Aufschlußverfahren für die Spurenanalytik. GIT Fachz. Lab. 2, 103–106 (1994).

Authors: *J. Begerow, L. Dunemann*
Examiner: *P. Schramel*

Monohydroxybutenyl-mercapturic acid (MHBMA) and dihydroxybutylmercapturic acid (DHBMA)

Application   Determination in urine

Analytical principle   High performance liquid chromatography/ tandem mass spectrometric detection (LC/MS/MS)

Summary

The procedure described here is suitable for the determination of monohydroxybutenylmercapturic acid (MHBMA) and dihydroxybutylmercapturic acid (DHBMA), two main metabolites of 1,3-butadiene, in the urine of persons exposed to the substance at the workplace or in the environment. After acidification of the urine, MHBMA and DHBMA are separated from interfering components of the matrix by means of a solid phase extraction. Then the analytes are separated by high performance liquid chromatography and quantified by tandem mass spectrometric detection. Ionisation is achieved by negative APCI (atmospheric pressure chemical ionisation). Calibration is carried out with spiked pooled urine samples from non-smokers in the concentration range from 0.5 µg/L to 500 µg/L (MHBMA) or 50.0 µg/L to 1250 µg/L (DHBMA). Deuterated structural analogues of the pure substances ($[D_6]$-MHBMA and $[D_7]$-DHBMA) are used as internal standards to enable quantification.

Monohydroxybutenylmercapturic acid (MHBMA)

Within-series imprecision: Standard deviation (rel.) $s_w = 4.1\%$
                                  Prognostic range $u = 10.5\%$
                                  at a spiked concentration of 7.35 µg MHBMA
                                  per litre urine and where n = 5 determinations

Between-day imprecision: Standard deviation (rel.) $s_w = 7.5\%$
                               Prognostic range $u = 15.7\%$
                               at a spiked concentration of 37.64 µg MHBMA
                               per litre urine and where n = 21 determinations

Accuracy:	Recovery rate and	$r = 93.7\%$ at 0.5 µg/L
		95.2% at 2 µg/L

Quantitation limit: 2.73 µg MHBMA per litre urine

Dihydroxybutylmercapturic acid (DHBMA)

Within-series imprecision: Standard deviation (rel.) $s_w = 0.8\%$
Prognostic range $u = 2.1\%$
at a spiked concentration of 496.5 µg DHBMA
per litre urine and where $n = 5$ determinations

Between-day imprecision: Standard deviation (rel.) $s_w = 5.7\%$
Prognostic range $u = 11.9\%$
at a spiked concentration of 394.8 µg DHBMA
per litre urine and where $n = 21$ determinations

Accuracy: Recovery rate $r = 95.3\%$ at 50 µg/L
and   73.6% at 1000 µg/L

Quantitation limit: 75.9 µg DHBMA per litre urine

1,3-Butadiene

H_2C⟶⟵CH_2

1,3-Butadiene is one of the most important basic chemicals. It is used alone for the manufacture of rubber and as a copolymer together with styrene [1]. The annual production is more than 5 million tonnes worldwide. Automobile exhaust gases [2] and the burning of fossil fuels (especially in domestic heating) represent important sources of environmental and human exposure. A car emits approx. 6 mg of 1,3-butadiene for every kilometre it travels [1]. Heating exhaust gases contain 33 mg/m^3. The half-life of 1,3-butadiene in the external air is estimated as approx. 4 h [3]. The concentration in urban air is 1 to 20 µg/m^3, depending on the density of traffic [4, 5]. Tobacco smoke is an important source of 1,3-butadiene in the air in enclosed spaces. Concentrations of between 2.7 and 19 µg/m^3 have been reported for 1,3-butadiene in smoke-filled rooms [1, 6, 7]. It was determined that smokers inhale 16 to 75 µg of 1,3-butadiene per cigarette in the mainstream smoke [7]. Exposure to 1,3-butadiene in food (values in the lower ppb range) and drinking water is thought to play a subordinate role [5, 8].
The International Agency for Research on Cancer (IARC) has classified 1,3-butadiene as "probably carcinogenic in humans" (Group 2A) [1]. The Commission has

Monohydroxybutenylmercapturic acid and dihydroxybutylmercapturic acid

assigned 1,3-butadiene to Carcinogen Category 1 ("causing cancer in humans") [9, 10]. A comprehensive review of the toxicological aspects of 1,3-butadiene can be found in the MAK value documentation [10].

Figure 1 shows an overview of the metabolism of 1,3-butadiene [11]. The first step of the metabolism of 1,3-butadiene is oxidation to the corresponding epoxides (1,2-epoxy-3-butene and 1,2,3,4-diepoxybutane) and unsaturated aldehydes (3-butenal, crotonaldehyde). In principle, the reactive epoxides can react with the cellular macromolecules, such as proteins or DNA, to form adducts or can be metabolised to hydroxymetabolites by epoxide hydrolases. An important metabolic pathway of the reactive intermediate products of 1,3-butadiene is conjugation with glutathione (GSH) (which may proceed under the enzymatic control of glutathione-S-transferases (GST) or spontaneously). These conjugates are further transformed to mercapturic acids. To date, three mercapturic acids have been identified for 1,3-butadiene: firstly, monohydroxybutenylmercapturic acid (MHBMA, sometimes designated MII in the literature), which consists of an isomeric mixture of 1-hydroxy-2-(N-acetylcysteinyl)-3-butene and 1-(N-Acetylcysteinyl)-2-hydroxy-3-butene, secondly, dihydroxybutylmercap-

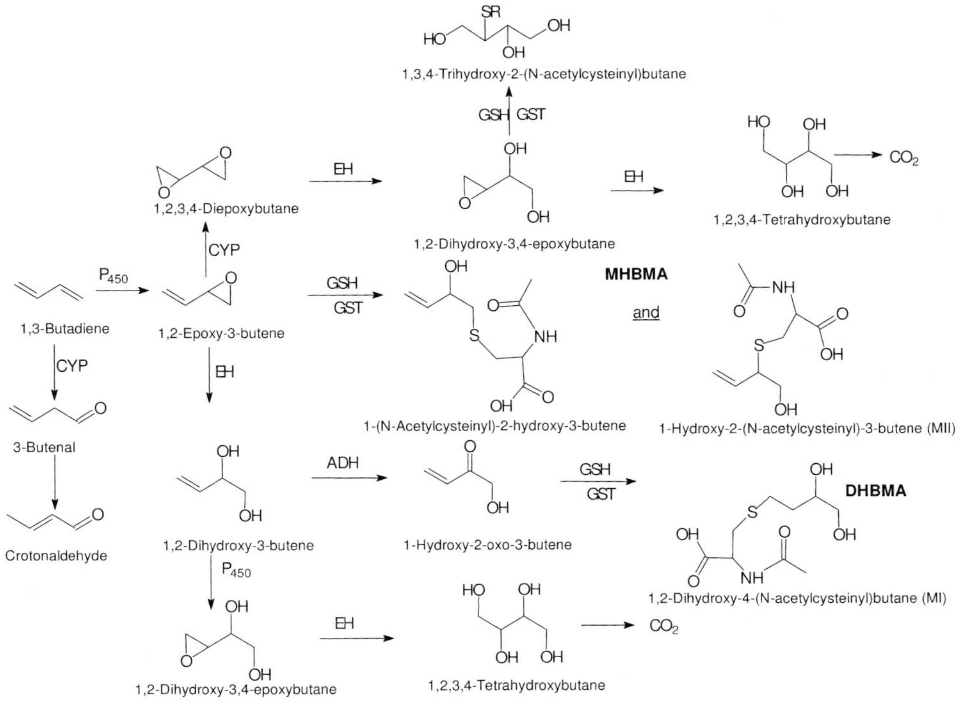

Fig. 1. Overview of the metabolism of 1,3-butadiene [11]. (SR = N-acetylcysteinyl group; GST = glutathione-S-transferase; GSH = glutathione; EH = epoxide hydrolase; CYP = cytochrome P_{450}, ADH = alcohol dehydrogenase)

turic acid (DHBMA or MI, systematic name 1,2-dihydroxy-4-(N-acetylcysteinyl)butane) and thirdly, 1,3,4-trihydroxy-2-(N-acetylcysteinyl)butane.

As a rule, the MHBMA and DHBMA metabolites are determined in urine for the purpose of biomonitoring. It can be assumed that DHBMA indicates the extent of the hydrolysis of 1,2-epoxy-3-butene to 1,2-dihydroxy-3-butene before the latter conjugates with GSH (probably after oxidation to unsaturated 1-hydroxy-2-oxo-3-butene). In contrast, MHBMA indicates the direct detoxification of 1,2-epoxy-3-butene by conjugation with GSH [12, 13]. The ratio of the metabolites in urine after exposure to 1,3-butadiene DHBMA/(DHBMA+MHBMA) is clearly dependent on the species and is approx. 0.9 for humans and monkeys [12–15], 0.25 to 0.5 for rats and 0.2 to 0.4 for mice [12, 14]. As 1,3-butadiene exposure increases the metabolic ratio decreases in various species [11, 14, 15].

Table 1 presents a brief overview of the MHBMA and DHBMA concentrations detected in investigations carried out in the fields of occupational and environmental medicine.

Table 1. MHBMA and DHBMA concentrations in the urine of persons after occupational and environmental exposure to 1,3-butadiene

Group	n	Concentration MHBMA [µg/L]	Concentration DHBMA [µg/L]	Reference
Investigations in occupational medicine				
Highly exposed workers	7	–	Mean 3200±1600	[14]
Moderately exposed workers	3	–	Mean 1390±550	[14]
Non-exposed Workers	10	–	Mean 630±190	[14]
Control group	9	–	Mean 320±70	[14]
Styrene-butadiene rubber workers	30	Range <0.1–962	Range 60–26207	[11]
1,3-Butadiene monomer workers	23	Range <0.1–44	Range 52–3522	[11]
Non-exposed control subjects	24	Range <0.1–8.2	Range 197–1211	[11]
Investigations in environmental medicine				
Smokers	10	Mean 86.4±14.0	Mean 644±90	[15]
Non-smokers	10	Mean 12.5±1.0	Mean 459±72	[15]

Authors: *G. Scherer, M. Urban*
Examiner: *W. Völkel*

Monohydroxybutenylmercapturic acid (MHBMA) and dihydroxybutylmercapturic acid (DHBMA)

Application	Determination in urine
Analytical principle	High performance liquid chromatography/ tandem mass spectrometric detection (LC/MS/MS)

Contents

1   General principles
2   Equipment, chemicals and solutions
2.1   Equipment
2.2   Chemicals
2.3   Solutions
2.4   Calibration standards
2.4.1 Internal standard
2.4.2 Calibration standards
3   Specimen collection and sample preparation
3.1   Sample preparation
4   Operational parameters
4.1   Operational parameters for high performance liquid chromatography
4.2   Operational parameters for mass spectrometry
5   Analytical determination
6   Calibration
7   Calculation of the analytical result
8   Standardisation and quality control
9   Evaluation of the method
9.1   Precision
9.2   Accuracy
9.3   Detection limits
9.4   Sources of error
10   Discussion of the method
11   References

The MAK-Collection Part IV: Biomonitoring Methods, Vol. 11.
DFG, Deutsche Forschungsgemeinschaft
Copyright © 2008 WILEY-VCH Verlag GmbH & Co. KGaA, Weinheim
ISBN: 978-3-527-31596-3

Monohydroxybutenylmercapturic acid and dihydroxybutylmercapturic acid

1 General principles

After acidification of the urine, monohydroxybutenylmercapturic acid and dihydroxybutylmercapturic acid are separated from interfering components of the matrix by means of a solid phase extraction. Then the analytes are separated by high performance liquid chromatography and quantified by tandem mass spectrometric detection. Ionisation is achieved by means of negative APCI. Calibration is performed using calibration standard solutions that are prepared in pooled urine from non-smokers and are treated in the same manner as the samples to be analysed. Deuterated structural analogues of the pure substances ($[D_6]$-MHBMA and $[D_7]$-DHBMA) are used as internal standards for quantification.

2 Equipment, chemicals and solutions

2.1 Equipment

LC/MS/MS system consisting of a binary high-pressure pump, column oven with a switch valve, degasser, automatic thermostatically controlled liquid sampling device and a tandem mass spectrometric detector with APCI, as well as a PC system for data evaluation.

Compressor (e.g. Manglitz model 4000 KCT-401-100-M.H.)

Membrane air dryer (e.g. Whatman model 64-01)

Nitrogen generator (e.g. Whatman model 75-72)

HPLC column:
Atlantis dC_{18}, length: 150 mm; inner diameter: 4.6 mm; particle diameter: 3 μm (e.g. Waters) with C_{18} pre-column, length: 4 mm; inner diameter: 3 mm, and precolumn holder (e.g. Phenomenex)

Test-tube shaker (e.g. Vortex, Carl Roth)

Variably adjustable pipettes: 10 to 100 μL, 100 to 1000 μL, 1000 to 5000 μL (e.g. Eppendorf Varipettes)

Volumetric flasks: 10 mL, 50 mL, 100 mL, 250 mL, 1000 mL

Glass beakers: 20 mL, 150 mL

100 mL Measuring cylinders

4 mL Graduated glass bottles with screw cap (e.g. Zefa Laborservice)

25 mL Amber glass bottles

0.5 to 2 L Polyethylene bottles for collecting human urine samples

10 mL Polyethylene tubes for portioning

133 Monohydroxybutenylmercapturic acid and dihydroxybutylmercapturic acid

Rotation evaporator (e.g. Büchi Labortechnik Rotavapor R-134)

Vacuum pump (e.g. Büchi Labortechnik)

Solid phase extraction columns (e.g. Waters Oasis HLB, 500 mg, 6 mL, No. 186000115)

Solid phase extraction station (e.g. Separtis VacMaster 20 with collection tube rack and lid 121-2016 and Isolute PTFE needle and stopcock, No. 121-0001)

Vacuum centrifuge (e.g. Jouan Speedvac Concentrator)

Roller mixer

Filter system 0.22 µm (e.g. Millipore GS)

100 mL Microvials (e.g. from Agilent Technologies)

Pipettes: 20 mL, 50 mL, 100 mL (e.g. from Brand)

pH meter (e.g. Lab 850, Schott Instruments)

Gassing system with four 250 mL washing bottles

Device for evaporation under a stream of nitrogen (e.g. Reacti-Therm, Pearce)

2.2 Chemicals

All chemicals must be analytical grade (p.a.) or purer.

(R)/(S)-N-acetyl-S-(1-hydroxymethyl)-2-propenyl)-L-cysteine / (R)/(S)-N-acetyl-S-(2-hydroxy)-3-butenyl)-L-cysteine, MHBMA (e.g. Toronto Research Chemicals No. A179005)

(R)/(S)-N-acetyl-S-(1-hydroxymethyl)-2-propenyl)-L-cysteine-D_6/(R)/(S)-N-acetyl-S-(2-hydroxy)-3-butenyl)-L-cysteine-D_6, [D_6]-MHBMA (e.g. Toronto Research Chemicals No. A179007)

N-acetyl-S-(3,4-dihydroxybutyl)-L-cysteine, DHBMA (e.g. Toronto Research Chemicals No. A173710)

N-acetyl-S-(3,4-dihydroxybutyl)-L-cysteine-D_7, [D_7]-DHBMA (e.g. Toronto Research Chemicals No. A173712)

Hydrochloric acid, 32% (e.g. Roth, Order No. P 074.1)

Formic acid, p.a., 98 to 100% (e.g. Merck No. 1.00264.1000)

Ammonium acetate p.a. (e.g. Merck No. 1.01116.0500)

Ammonia 3.8, anhydrous (e.g. Linde No 4930795)

Nitrogen 5.0 (e.g. from Linde)

Deionised water (e.g. Seradest ultrapure water plant)

Methanol, HPLC grade (e.g. Promochem code 3041)

Ethyl acetate, trace analysis grade (e.g. Promochem code 1191.1)

Acetonitrile, HPLC grade (e.g. Merck No. 1.00030.9010)

2.3 Solutions

4 M Hydrochloric acid:
Approx. 100 mL water are placed in a 250 mL volumetric flask and a pipette is used to carefully add 100 mL hydrochloric acid (32%). After the concentrated hydrochloric acid has been pipetted, the pipette is rinsed several times with the contents of the flask to remove any residual hydrogen chloride gas from the body of the pipette. The flask is subsequently filled to its nominal volume with water. The solution is kept at room temperature and can be used over a period of 6 months.

1 M Hydrochloric acid:
Using a pipette, 50 mL hydrochloric acid (32%) are added to a 500 mL volumetric flask into which approx. 250 mL water has been previously placed. After the concentrated hydrochloric acid has been pipetted, the pipette is rinsed several times with the contents of the flask to remove any residual hydrogen chloride gas from the body of the pipette. The flask is subsequently filled to its nominal volume with water. The solution is kept in the refrigerator at 4 to 6 °C and can be used over a period of 6 months.

2% Formic acid pH 2 for HPLC:
Approx. 500 mL deionised water are placed in a 1000 mL volumetric flask and a pipette is used to add 20 mL formic acid. The flask is subsequently filled to its nominal volume with water. The solution is filtered through a 0.22 μm filter.
The solution is stored at 4 to 6 °C in the refrigerator. It is stable for 1 week.

Hydrochloric acid pH 2.0:
1 M hydrochloric acid is added drop by drop using a pipette to 1000 mL deionised water in a glass beaker, while the pH is constantly checked with a pH meter, until the desired pH value is reached. The solution is stored at room temperature in a sealable glass vessel. It is stable for 1 week.

Ethyl acetate (saturated with ammonia) with 20% MeOH (v/v):
A washing bottle filled with 250 mL ethyl acetate is integrated in second place into a gassing system that consists of four washing bottles. The first and third bottles are empty. The fourth bottle is two-thirds filled with water. All the bottles are secured by clamps. The first bottle is connected to the pressure valve of an ammonia bottle. The valve is opened for 3 min. 50 mL methanol are placed in a 250 mL volumetric flask and the flask is filled to its nominal volume with the freshly prepared ethyl acetate solution.

135 Monohydroxybutenylmercapturic acid and dihydroxybutylmercapturic acid

The solution is kept at room temperature and can be used over a period of 1 week.

0.5% Acetonitrile in hydrochloric acid pH 2.0:
0.5 mL Acetonitrile are placed in a 100 mL volumetric flask. The flask is subsequently filled to its nominal volume with hydrochloric acid pH 2.0. The solution is stored at room temperature. The solution is stable for 1 week.

2.4 Calibration standards

2.4.1 Internal standard

$[D_6]$-MHBMA stock solution:
10.0 mg of (R)/(S)-N-acetyl-S-(1-hydroxymethyl)-2-propenyl)-L-cysteine-D_6/(R)/(S)-N-acetyl-S-(2-hydroxy)-3-butenyl)-L-cysteine-D_6 are weighed exactly in a 10 mL volumetric flask. The flask is subsequently filled to its nominal volume with methanol (1 g/L).

$[D_7]$-DHBMA stock solution:
10.0 mg of N-acetyl-S-(3,4-dihydroxy-butyl)-L-cysteine-D_7 are weighed exactly in a 10 mL volumetric flask. The flask is subsequently filled to its nominal volume with methanol (1 g/L).

$[D_6]$-MHBMA working solution ISTD I:
100 µL of the $[D_6]$-MHBMA stock solution are pipetted into a 10 mL volumetric flask. The flask is subsequently filled to its nominal volume with methanol (10 mg/L).

$[D_7]$-DHBMA working solution ISTD II:
100 µL of the $[D_7]$-DHBMA stock solution are pipetted into a 10 mL volumetric flask. The flask is subsequently filled to its nominal volume with methanol (10 mg/L).

The stock solutions and working solutions of the internal standards are frozen in 25 mL amber glass bottles at –18 °C and are stable for at least 12 months under these conditions.

2.4.2 Calibration standards

MHBMA stock solution:
10.0 mg of (R)/(S)-N-acetyl-S-(1-hydroxymethyl)-2-propenyl)-L-cysteine / (R)/(S)-N-acetyl-S-(2-hydroxy)-3-butenyl)-L-cysteine are weighed exactly in a 10 mL volumetric flask. The flask is subsequently filled to its nominal volume with methanol (1 g/L).

DHBMA stock solution:
10.0 mg of N-acetyl-S-(3,4-dihydroxy-butyl)-L-cysteine are weighed exactly in a 10 mL volumetric flask. The flask is subsequently filled to its nominal volume with methanol (1 g/L).

MHBMA working solution I:
100 µL of the MHBMA stock solution are pipetted into a 10 mL volumetric flask. The flask is subsequently filled to its nominal volume with methanol (10 mg/L).

MHBMA working solution II:
100 µL of MHBMA working solution I are pipetted into a 10 mL volumetric flask. The flask is subsequently filled to its nominal volume with methanol (100 µg/L).

DHBMA working solution:
100 µL of the DHBMA stock solution are pipetted into a 10 mL volumetric flask. The flask is subsequently filled to its nominal volume with methanol (10 mg/L).

The stock solutions and working solutions of the calibration standards are frozen in 25 mL amber glass bottles at −18 °C and are stable for at least 12 months under these conditions.

The calibration standard solutions are prepared as follows:
Frozen pooled human urine is thawed overnight. As soon as it has reached room temperature, it is shaken vigorously and then centrifuged (10 °C, 2000 g, 10 min). A 20 mL aliquot of the supernatant liquid is placed in a glass beaker and the pH is adjusted initially to pH 3.0–2.5 by adding 4 M hydrochloric acid drop by drop using a pH meter while the liquid is stirred constantly, and then the pH is adjusted to exactly 2.0 using 1 M hydrochloric acid (see also Section 3.1). The volumes of MHBMA working solutions I and II, the DHBMA working solution and the solutions of the internal standards ISTD I and II shown in Table 2 are each pipetted into a 10 mL PE tube, and are then evaporated to dryness under a gentle stream of nitrogen without heating. Then 2 mL of the acidified pooled urine are pipetted into each tube and the contents are mixed on a roller mixer for at least 10 min in order to ensure complete dissolution of the standard substances in urine (see pipetting scheme in Table 2).

Table 2. Pipetting scheme for the preparation of the calibration standard solutions in human urine

Volume of the working solutions			Volume of the spiking solutions of the internal standards		Volume of the pooled urine (adjusted to pH 2.0)	Concentration of the calibration standards	
MHBMA		DHBMA	$[D_6]$-MHBMA	$[D_7]$-DHBMA		MHBMA	DHBMA
I [µL]	II [µL]	[µL]	ISTD I [µL]	ISTD II [µL]	[mL]	[µg/L]	[µg/L]
–	25	25	100	100	2	1.25	125
–	50	50	100	100	2	2.5	250
–	100	100	100	100	2	5.0	500
–	250	250	100	100	2	12.5	1250
10	–	500	100	100	2	50	2500
25	–	750	100	100	2	125	3750
100	–	–	100	100	2	500	–
250	–	–	100	100	2	1250	–

137 Monohydroxybutenylmercapturic acid and dihydroxybutylmercapturic acid

The resulting calibration standard solutions are processed as described in Section 3.1, whereby the working steps before the solid phase extraction are omitted for the calibration standard solutions, as they are used directly for the solid phase extraction.

3 Specimen collection and sample preparation

The polyethylene bottles used to collect the urine specimens are washed with a cleansing agent and rinsed with bidistilled water. The samples are divided into aliquots after collection (5 to 10 mL sample in 10 mL tubes) and they are then deep-frozen (–18 °C) until analysis is carried out. Stabilisation is unnecessary. The aliquots can be used for at least 6 months.

3.1 Sample preparation

The deep-frozen urine samples are thawed overnight. As soon as they have reached room temperature, they are shaken vigorously and then centrifuged (10 °C, 2000 g, 10 min). The supernatant liquid is cautiously withdrawn, transferred to a new 10 mL tube and the pH value is initially adjusted to 3.0–2.5 using 4 M hydrochloric acid while the contents are stirred continuously, and then the pH is adjusted to exactly 2.0 using 1 M hydrochloric acid. The electrode of the pH meter must be immersed in the sample tube during the entire pH adjustment. Adjustment of the pH value is regarded as complete when the pH value does not change over a period of at least 20 seconds without the contents being shaken. To perform a duplicate determination of human urine, 100 µL [D_6]-MHBMA working solution ISTD I (10 mg/L) and 100 µL [D_7]-DHBMA working solution ISTD II (10 mg/L) are each pipetted into a 10 mL PE tube, and are then evaporated to dryness under a gentle stream of nitrogen without heating. Then 2 mL of the centrifuged human urine are pipetted into each tube, and the contents are mixed on a roller mixer for at least 10 min in order to ensure complete dissolution of both internal standards in urine.

Then solid phase extraction is carried out on the Oasis HLB polymer material (500 mg cartridge, 6 mL). First the cartridges are conditioned with 2×6 mL ethyl acetate, 2×6 mL methanol and 2×6 mL dilute hydrochloric acid pH 2.0. After complete application of the samples at atmospheric pressure, the cartridges are washed first with 3×3 mL dilute hydrochloric acid pH 2.0, then with 6 mL of the solution of 0.5% acetonitrile in dilute hydrochloric acid pH 2.0, and then dried under vacuum (550 mbar, 3 min). Subsequently the cartridges are centrifuged (10 °C, 500 g, 10 min) and dried in a stream of nitrogen with the help of the drying lid. The prepressure at the nitrogen bottle is set at 2.0 bar for this purpose. When the aqueous phase has been completely compressed from the cartridge, drying is continued for 2 min.

Elution is carried out using the solution of 20% methanol in ethyl acetate saturated with ammonia (v/v) into a 4 mL vial. For this purpose 5 mL of the solution are applied and, after being allowed to act for 1 min, the solution is passed through under slightly reduced pressure at 920 mbar so that drop-by-drop elution can be observed. In order to achieve complete elution the cartridges are subjected to a vacuum of 550 mbar for 2 min until complete dryness. The sample is evaporated to dryness using a vacuum centrifuge and then taken up in 100 µL of a 2% formic acid/methanol mixture 70 : 30 (v/v). The analytical determination is subsequently carried out by means of LC/MS/MS.

Figure 2 shows the sample processing in the form of a flowchart.

4 Operational parameters

The analytical measurements are performed with a combination of instruments comprising a HPLC system with a binary pump, column oven, degasser and autosampler, as well as a tandem quadrupole mass spectrometer with the possibility of negative APCI.

4.1 Operational parameters for high performance liquid chromatography

Separation column:	Material:	Steel
	Length:	150 mm
	Inner diameter:	4.6 mm
	Column packing:	Atlantis dC_{18}, 3 µm, 100 Å
Separation principle:	Reversed phase	
Temperature:	50 °C	
Detection:	Tandem mass spectrometric detector	
Mobile phase:	Eluent A:	2% Formic acid pH 2.0, aq.
	Eluent B:	Methanol
Gradient:	See Table 3	

Table 3. Gradient program of the binary pump

Time (min)	Eluent A vol. %	Eluent B vol. %
0.00	70	30
1.00	70	30
4.00	10	90
5.00	10	90
5.01	70	30
10.00	70	30

Monohydroxybutenylmercapturic acid and dihydroxybutylmercapturic acid

Stop time: 10 minutes

Flow rate: 1 mL/min

Autosampler: Cooling: 10 °C
Injection volume: 10 µL

All other parameters must be optimised in accordance with the manufacturer's instructions.

4.2 Operational parameters for mass spectrometry

Settings for the ion source:

Ionisation mode: APCI negative

Source temperature: 495 °C

Curtain gas pressure: 50 psi

Needle current (NC): −2 µA

Nebulizer gas (GAS 1): 70 psi

Auxiliary/heater gas (GAS 2): 20 psi

Settings for the analyser: See Table 4

Table 4. Analyser settings

Parameter Transition	$[D_7]$-DHBMA 257.2 → 128.0	DHBMA 250.0 → 121.0	250.0 → 75.0	$[D_6]$-MHBMA 238.2 → 109.0	MHBMA 232.0 → 103.0	232.0 → 73.0
Measuring time [ms]	200	200	200	200	200	200
DP [V]	−31	−56	−56	−26	−31	−31
FP [V]	−80	−190	−190	−60	−330	−330
EP [V]	9.5	8.0	8.0	10.0	6.0	6.0
CEP [V]	−8.0	−52	−52	−10	−10	−10
CXP [V]	−8	−6	−14	−8	−24	−8
CE [V]	−20	−22	−36	−12	−14	−34

DP = declustering potential, FP = focussing potential, EP = entrance potential, CEP = collision cell entrance potential, CXP = collision cell exit potential, CE = collision energy.

The measurement conditions listed here were established for the configuration of instruments used in this case and they must be optimised for the instruments of other manufacturers in accordance with their instructions.

5 Analytical determination

For quantitative determination, 10 µL of the sample solution are injected onto the separation column. Two quality control samples of different concentrations are included in each analytical series of 18 samples. A reagent blank value is determined every day. Each real sample is assayed in duplicate. If the resulting measurement values are outside the linear range of the calibration curve, the urine is diluted appropriately with deionised water (e.g. 1 : 10) before processing and measuring the samples anew. The ion transitions shown in Table 5 are recorded in the MRM mode of the tandem mass spectrometer (APCI negative mode).

Table 5. Retention times and detected ion transitions

Analyte	Retention time [min]	Ion transitions (MS/MS, APCI neg. mode)	
		Q 1	Q 3
[D_7]-DHBMA	2.19	257.2	128.0*
DHBMA	2.20	250.0	121.0*
			75.0
[D_6]-MHBMA	2.97	238.2	109.0*
MHBMA	3.01	232.0	103.0*
			73.0

The masses marked * are used for quantitative evaluation.

The retention times shown in Table 5 serve only as a guide. Users of the method must satisfy themselves of the separation power of the HPLC column they use and of the resulting retention behaviour of the substances.

Figures 3a and 3b show examples of the chromatograms of two urine samples (a smoker and a non-smoker).

6 Calibration

A matrix calibration was carried out using the control urine spiked as shown in Table 2 and processed as described in Section 3.1, and using the instrumental parameters listed in Sections 4.1 and 4.2. The calibration curve is plotted by linear regression of the area ratios of MHBMA/[D_6]-MHBMA or DHBMA/[D_7]-DHBMA as a function of the spiked concentrations of MHBMA or DHBMA, respectively. The concentration without spiking is subtracted from the values in each case. When a sample of 2 mL was used, a linear measurement range from 1.25 to 1250 µg/L (for MHBMA) and from 125 to 3750 µg/L (for DHBMA) was achieved with the analyti-

cal instruments described here. It is not necessary to plot a complete calibration graph for every analytical series. New calibration graphs should be plotted only if the quality assurance results indicate systematic deviations, changes have been made to the system, or no analysis has been carried out for a longer period of time.
Examples of linear calibration graphs are shown in Figure 4.

7 Calculation of the analytical result

The concentration of the sample in µg/L can be calculated from the calibration function ($y = mx + t$) according to the following equation:

Concentration of the sample [µ/L] = $(y - t)/m$

where y: Peak ratio: analyte/ISTD
　　　 t: Intercept with the y-axis of the calibration function after the blank value has been subtracted (see Section 6)
　　　 m: Slope of the calibration function

Any reagent blank values must be subtracted from the analytical results for the real samples. This calculation may also be carried out by the evaluation software of the LC/MS/MS system (e.g. "Analyst" software from Applied Biosystems), if appropriate. If a sample has been diluted before processing, the concentration calculated from the above equation must be multiplied by the dilution factor.

8 Standardisation and quality control

Quality control of the analytical results is carried out as stipulated in the guidelines of the Bundesärztekammer (German Medical Association) [16, 17] and in the special preliminary remarks to this series. Pooled urine samples from smokers and non-smokers that were prepared in the laboratory are included in the analysis to check the reproducibility (precision from day to day). The "expected value" and the tolerance range of this quality control material are determined in a preliminary study (precision in the series with $n = 5$). A reagent blank value is determined every day. Two control samples with 2 different concentrations (e.g. pooled smoker sample, pooled non-smoker sample) are included in each processing series of 18 real samples in each case in order to record system changes. One control sample is measured before the 18 real samples and the second control is measured after the 18 samples. One standard solution of the analytes is measured per 100 injections. This standard sample is equivalent to a mixture of the 4 individual standards, the concentration of the individual components being 2.5 mg/L. If the results for the quality control samples indicate systematic deviations, calibration as described in Section 6 must be carried out anew.

9 Evaluation of the method

9.1 Precision

The precision in the series and from day to day was determined using unspiked human urine. One sample of urine from a smoker and one from a non-smoker were processed and analysed 5 times as described in the previous sections to check the precision in the series. The precision from day to day was determined as described in the previous sections by processing and analysing one aliquot of a pooled urine sample from smokers and one from non-smokers once a day on 21 days over a period of 5 weeks.
The results are listed in Table 6.

Table 6. Precision for MHBMA and DHBMA determinations in human urine

	n	Concentration [µg/L]	Variation coefficient [%]	Prognostic range [%]
MHBMA				
Non-smokers				
In the series	5	7.35	4.1	10.5
From day to day	21	37.6	7.5	15.7
Smokers				
In the series	5	9.35	6.3	16.2
From day to day	21	18.4	13.4	27.9
DHBMA				
Non-smokers				
In the series	5	496.5	0.8	2.1
From day to day	21	394.8	5.7	11.9
Smokers				
In the series	6	238.1	1.0	2.6
From day to day	21	219.8	4.2	8.7

9.2 Accuracy

The accuracy was checked for human urine by means of experiments with spiked pooled urine samples from non-smokers. For this purpose 5 pooled samples were each spiked with 2.5 and 10 ng MHBMA in a batch of 5 mL. The spiked amounts of DHBMA were 250 and 5000 ng. These samples were subsequently processed and analysed in accordance with Sections 3 and 4. The results are shown in Table 7.

143 Monohydroxybutenylmercapturic acid and dihydroxybutylmercapturic acid

Table 7. Accuracy for MHBMA and DHBMA determinations in human urine (n=5, starting concentration; 4.27 µg/L and 476.4/L respectively)

Spiked Concentration [µg/L]	Relative recovery rate [%] (Variation coefficient [%])	
	MHBMA	DHBMA
0.5	93.7 (8.1)	–
2.0	95.2 (6.1)	–
50	–	95.3 (4.6)
1000	–	73.6 (2.9)

9.3 Detection limits

The detection and quantitation limits were calculated from the signal/background noise ratios from 10 pooled urine samples of non-smokers based on the following ratios:

Detection limit $(\underline{X}*) = 3\,\sigma_{blank}$

Quantitation limit $= 10\,\sigma_{blank}$

whereby σ_{blank} is the standard deviation of the mean blank value at the retention time of the mercapturic acid in question.
The results are shown in Table 8.

Table 8. Detection and quantitation limits of MHBMA and DHBMA in human urine

Analyte	DL [µg/L]	QL [µg/L]
MHBMA	0.91	2.73
DHBMA	23.0	75.9

DL = Detection Limit, QL = Quantitation Limit.

9.4 Sources of error

The chromatographic separation of the analytes by the gradient program is not only necessary due to the differing polarity of the analytes, but above all it also serves to separate interfering matrix components. The use of the strongly acidic HPLC solvent mixture permits a relatively good retention on the C_{18} column with 3 µm particles, so that the peaks can be clearly assigned by comparing the peak forms and retention times of the native metabolites with those of the internal standards. The isomeric mixture of MHBMA is not completely separated by means of this chromatography (see Figures 3a and 3b). This method evaluates the sum of the two MHBMA isomers.

High specificity is ensured by detection of daughter ions by means of LC/MS/MS. The deuterated standards and their retention times (immediately before the analytes) may also serve as further identification criteria.

The drying of the SPE cartridges before elution can be regarded as a critical step. Incomplete drying may lead to uncontrolled losses of the analytes. Therefore it is essential to ensure complete drying of the SPE cartridges.

Our experience to date shows that MHBMA and DHBMA dissolved in urine remain stable for at least 12 months when stored at $-25\,°C$.

10 Discussion of the method

This newly developed method to determine the mercapturic acids of 1,3-butadiene permits rapid, highly selective and sensitive quantification of 2 mercapturic acids simultaneously by means of HPLC/APCI-MS/MS with a measurement duration of less than 10 minutes. Thus a method of monitoring these classical detoxification products has been made available that permits the throughput of a relatively high number of samples in a short time. Up to 60 samples per week can be assayed as duplicates on account of the simple sample preparation. The principle of the method is a simple solid phase extraction on a polymer material. The adsorption of the analytes on the polymer material can only be based on $p\pi$ or $p\pi^*$ interactions ("normal" reversed phase chromatography). Therefore the analytes, due to their high polarity, should be relatively loosely and in particular non-specifically bound. The unchanging flow rates required for constant recovery can be achieved for manual solid phase extraction only by means of a vacuum controller. It was shown that the mercapturic acids have sufficient thermal stability for measurement with APCI. Temperatures of $500\,°C$ in the ion source are necessary due to the high proportion of water in the eluent. Thus the analytes reach temperatures of approx. $200\,°C$ for brief periods of time. However, no measurable fragmentation occurs in the source. Measurement with APCI has some advantages over the TIS (turbo ion spray) source, as relatively short run times of 10 min can be achieved at higher flow rates of 1000 µL/min and column diameters of 4.6 mm, and the signal/background noise ratio is distinctly improved with the use of APCI. However, the absolute intensity of the signals is slightly reduced.

Both the calibration and the determination of the detection and quantitation limits were carried out using urine from non-smokers. As no urine is available that is free from the 1,3-butadiene mercapturic acids, this inevitably leads to an over-estimation of the detection and quantitation limits if they are determined by the blank value method. These are more objectively evaluated by the signal/background noise ratio.

The method is linear over 3 orders of magnitude for all the analytes. The selectivity is confirmed by recording full-scan mass spectra as part of the "quantitative optimisation".

In the course of the method development it was demonstrated that the amount of urine used has a definite influence on the recovery. When larger quantities of urine are used, there is a distinct decline in the recovery, which indicates that binding sites of the cartridges are mainly occupied by matrix components and therefore are no longer available for retention of the analytes.

Instruments used:
HP 1100 HPLC system from Agilent Technologies with G1312 A, G1314 A, G1316 A, G1322 A and G1329 A modules and a tandem quadrupole mass spectrometer with an API 2000 APCI source, as well as the "Analyst" evaluation software from Applied Biosystems.

11 References

[1] *International Agency for Research on Cancer (IARC):* 1,3-Butadiene. Monographs on the Evaluation of the Carcinogenic Risk of Chemicals to Humans: World Health Organization, Lyon (1992).
[2] *N. Pelz, A.M. Dempster* and *P.R. Shore:* Analysis of low molecular weight hydrocarbons including 1,3-butadiene in engine exhaust gases using an aluminum oxide porous-layer open-tubular fused-silica column. J. Chromatogr. Sci. 28, 230–235 (1990).
[3] *L.I. Cote* and *S.P. Bayard:* Cancer: Risk assessment of 1,3-Butadiene. Environ. Health Perspect. 86, 149–153 (1990).
[4] *L. Löfgren* and *G. Petersson:* Butenes and butadiene in urban air. Sci. Total Environ. 116, 195–201 (1992).
[5] *International Agency for Research on Cancer (IARC):* Some Chemicals used in plastics and elastomers. Monographs on the evaluation of the carcinogenic risk of chemicals to Humans. World Health Organisation, Lyon (1987).
[6] *G. Löfroth, R.M. Burton, L. Forehand, S.K. Hammond, R.L. Seila, R.B. Zweidinger* and *J. Lewtas:* Characterization of environmental tobacco smoke. Environ. Sci. Technol. 23, 610–614 (1989).
[7] *K.D. Brunnemann, M.R. Kagan, J.E. Cox* and *D. Hoffmann*: Analysis of 1,3-butadiene and other selected gas-phase components in cigarette mainstream and sidestream smoke by gas chromatography – mass selective detection. Carcinogenesis 11, 1863–1868 (1990).
[8] *T.P. McNeal* and *C.V. Breder:* Headspace gas chromatographic determination of residual 1,3-butadiene in rubber-modified plastics and its migration from plastic containers into selected foods. J. Assoc. Offic. Anal. Chem. 70, 18–21 (1987).
[9] *Deutsche Forschungsgemeinschaft:* List of MAK and BAT Values 2006, 42nd report, Wiley-VCH, Weinheim (2006).
[10] *H. Greim (ed.):* 1,3-Butadiene. Occupational Toxicants – Critical Data Evaluation for MAK Values and Classification of Carcinogens. Vol. 15, Wiley-VCH, Weinheim (2001).
[11] *N.J. van Sittert, H.J.J.J. Megens, W.P. Watson* and *P.J. Boogaard:* Biomarkers of exposure to 1,3-butadiene as a basis for cancer risk assessment. Toxicol. Sci. 56, 189–202 (2000).
[12] *P.J. Sabourin, L.T. Burka, W.E. Bechtold, A.R. Dahl, M.D. Hoover, I.Y. Chang* and *R.F. Henderson:* Species differences in urinary butadiene metabolites; identification of 1,2-dihydroxy-4-(N-acetylcysteinyl)butane, a novel metabolite of butadiene. Carcinogenesis 13, 1633–1638 (1992).
[13] *P.J. Boogaard, N.J. van Sittert* and *H.J.J.J. Megens:* Urinary metabolites and haemoglobin adducts as biomarkers of exposure to 1,3-butadiene: a basis for 1,3-butadiene cancer risk assessment. Chem. Biol. Interact. 135–136, 695–701 (2001).
[14] *W.E. Bechtold, M.R. Strunk, I.Y. Chang, J.B. Ward Jr.* and *R.F. Henderson:* Species differences in urinary butadiene metabolites: Comparisons of metabolite ratios between mice, rats and humans. Toxicol. Appl. Pharmacol. 127, 44–49 (1994).

[15] *M. Urban, G. Gilch, G. Schepers, E. Van Miert* and *G. Scherer:* Determination of the major mercapturic acids of 1,3-butadiene in human and rat urine using liquid chromatography-tandem mass spectrometry. J. Chrom. B 796, 131–140 (2003).

[16] *Bundesärztekammer:* Qualitätssicherung der quantitativen Bestimmungen im Laboratorium. Neue Richtlinien der Bundesärztekammer. Dt. Ärztebl. 85, A699–A712 (1988).

[17] *Bundesärztekammer:* Ergänzung der „Richtlinien der Bundesärztekammer zur Qualitätssicherung in medizinischen Laboratorien" Dt. Ärztebl. 91, C159–C161 (1994).

Authors: *G. Scherer, M. Urban*
Examiner: *W. Völkel*

Monohydroxybutenylmercapturic acid and dihydroxybutylmercapturic acid

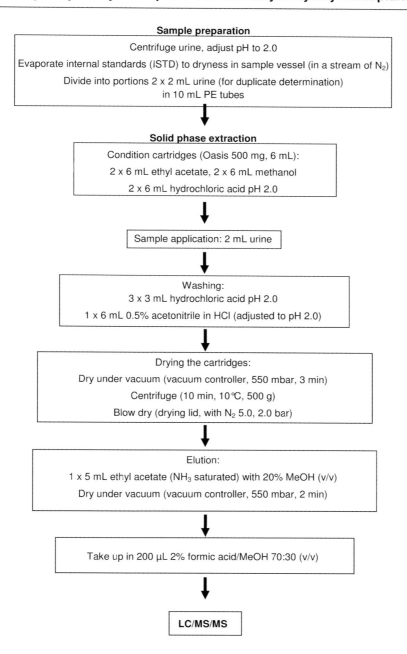

Fig. 2. Method flowchart

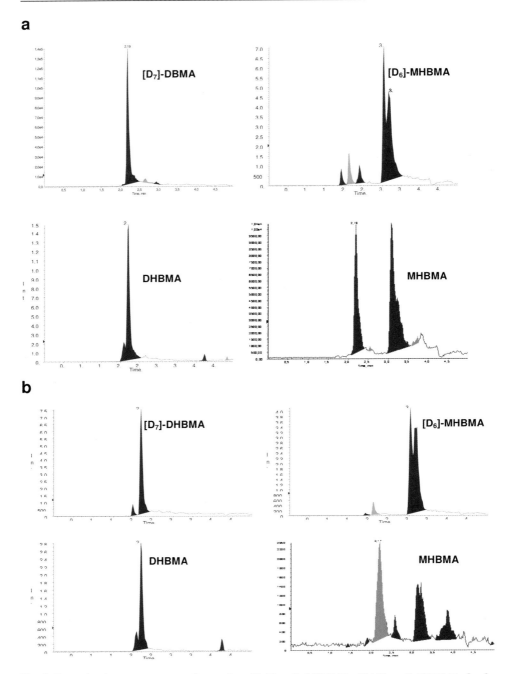

Fig. 3. Example chromatograms: **a** of a smoker: 11.43 µg/L MHBMA; 424.81 µg/L DHBMA. **b** of a non-smoker: 1.63 µg/L MHBMA; 112.3 µg/L DHBMA

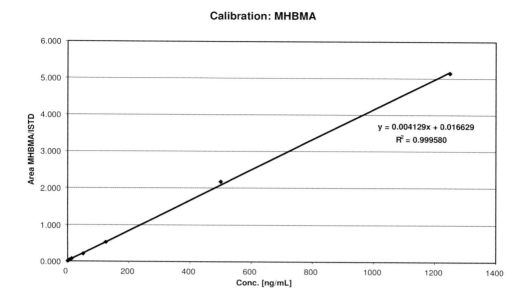

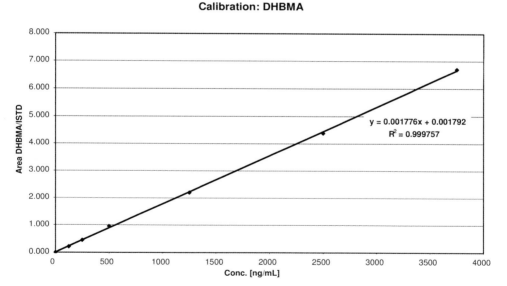

Fig. 4. Linear calibration graphs for MHBMA and DHBMA. A matrix calibration was carried out

Neuropathy target esterase (NTE)

Application Determination in leukocytes

Analytical principle Photometric determination at 492 nm

Summary

The method described here permits the determination of the activity of neuropathy target esterase (NTE) in isolated leukocytes following occupational exposure to organophosphates.

The leukocytes are previously isolated and subjected to haemolysis using deionised water. The protein concentration of the haemolysate must be adjusted to 200 µg/mL before determination of the NTE activity. For this purpose the protein content must be previously assayed by means of a Bradford test. The determination of the NTE activity is based on the principle of enzymatic cleavage of the substrate phenylvalerate to phenol, which undergoes a colour reaction that can be measured by means of photometry at 492 nm (see Figure 1).

Two organophosphates with different modes of action are added to the leukocyte sample in order to determine the absolute NTE activity. Firstly paraoxon, a non-spe-

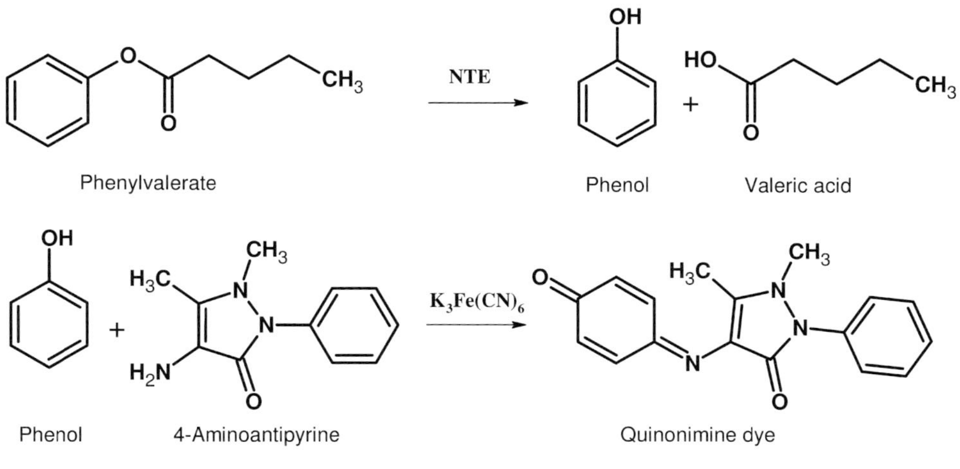

Fig. 1. Reactions on which the photometric determination of the NTE activity are based

cific esterase inhibitor that, however, does not inhibit the NTE activity, is added and secondly mipafox is added, which acts as a sensitive inhibitor in combination with paraoxon. The absolute NTE activity is calculated from the difference between the phenylvalerate concentrations in the leukocyte samples containing only paraoxon and in those samples that also contain mipafox.

Neuropathy target esterase activity

Within-series imprecision: Standard deviation (rel.) $s_w = 1.8\%$
(Phenol determination)   Prognostic range $u = 3.9\%$
                             at a spiked concentration of 40 mg phenol per litre
                             leukocyte haemolysate and where n = 10 determinations

Between-day imprecision: was not determined (see Section 8.1)

Accuracy: was not determined (see Section 8.2)

Detection limit: was not determined (see Section 8.3)

Neuropathy target esterase

Some organophosphates (OPs), such as chlorpyrifos, MCCP, mipafox, trichlorfon, TOCP, trichloronate, leptophos, EPN and methamidophos, may induce delayed neuropathy (organophosphorus induced delayed neuropathy, OPIDN). OPIDN belongs to the neurodegenerative disorders. Delayed neurotoxicity is defined as the retarded onset of prolonged locomotive ataxia as a consequence of a single or repeated exposure to OPs. Both the sensory and the motor nerve fibres in the central and peripheral nervous system are affected, resulting in distal neuropathy. The clinical manifestation of this disorder can appear with a delay of up to 4 weeks [1].

Animal studies have shown that irreversible inhibition of neuropathy target esterase (NTE) in the nervous tissue may be regarded as the first indicator of the onset of OPIDN. The NTE activity in the nervous tissue is correlated with that in lymphocytes (in humans too). However, there is no connection between AChE (acetylcholinesterase) inhibition and NTE inhibition. NTE activity is inhibited within hours following OP exposure. In contrast, lymphocytic NTE, with a regeneration half-life of 5 to 7 days, can already have regained its original pre-exposure activity when the first clinical symptoms appear [2].

The NTE activity was determined in 137 healthy workers. The activities were between 3.7 and 21.8 nmol phenylvalerate/min/mg protein. The mean value was 8.1 ± 3.0 (SD) and the median was 7.8 nmol/min/mg protein [3].

An investigation of 108 Caucasians showed a mean activity of 11.5 ± 2.5 nmol/min/mg protein. The mean intra-individual variation coefficient was 10.1%. Differences due to gender or age were not detectable [4].

The relationship between the development of OPIDN and the inhibition of lymphocytic NTE has not yet been investigated in man. In hens, an animal species that is sensitive to this effect, it was established that more than 70% (more likely 90%) of the NTE activity must be inhibited in order to induce OPIDN [5].

It is essential in any case to determine the NTE activity before OP exposure due to the intra-individual variation range of the NTE activity. Only the difference from the original findings (i.e. at least 70% reduction in the NTE activity) permits possible conclusions. Therefore at present NTE determination is only meaningful in occupational medicine as part of preventive examinations of workers who may be exposed to OPs, as values prior to exposure can be ascertained in that case.

The fact that only a few OPs inhibit NTE must also be taken into account. Therefore the AChE activity must also be determined in each case in order to evaluate the effects of handling OPs [6, 7]. OP intoxication due to an OP that leads to strong inhibition of AChE but only slight NTE inhibition can cause death due to internal cholinergic poisoning. This occurs before the dose necessary for development of OPIDN has been reached [2].

As no comprehensive field studies have been published to date, routine determination of the NTE activity is recommended, at least for those workers who handle OPIDN-inducing OPs.

Authors: *J. Lewalter, G. Leng, B. Röhrig*
Examiner: *M. Müller*

Neuropathy target esterase (NTE)

Application Determination in leukocytes

Analytical principle Photometric determination at 492 nm

Contents

1 General principles
2 Equipment, chemicals and solutions
2.1 Equipment
2.2 Chemicals
2.3 Solutions
2.4 Calibration standards
2.4.1 Standard solutions for protein assay
2.4.2 Standard solutions to determine the activity of NTE
3 Specimen collection and sample preparation
3.1 Sample preparation
4 Operational parameters
5 Analytical determination
5.1 Protein assay
5.2 NTE activity determination
6 Calibration
7 Calculation of the analytical result
8 Evaluation of the method
8.1 Precision
8.2 Accuracy
8.3 Detection limit
8.4 Sources of error
9 Discussion of the method
10 References

1 General principles

The leukocytes are previously isolated and subjected to haemolysis using deionised water. The protein concentration of the haemolysate must be adjusted to 200 µg/mL before determination of the NTE activity. For this purpose the protein content must be previously assayed by means of a Bradford test. The determination of the NTE activity is based on the principle of enzymatic cleavage of the substrate phenyl-

The MAK-Collection Part IV: Biomonitoring Methods, Vol. 11.
DFG, Deutsche Forschungsgemeinschaft
Copyright © 2008 WILEY-VCH Verlag GmbH & Co. KGaA, Weinheim
ISBN: 978-3-527-31596-3

valerate to phenol, which undergoes a colour reaction that can be measured by means of photometry at 492 nm (see Figure 1).

Two organophosphates with different modes of action are added to the leukocyte sample in order to determine the absolute NTE activity. Firstly paraoxon, a non-specific esterase inhibitor that, however, does not inhibit the NTE activity, is added and secondly mipafox is added, which acts as a sensitive inhibitor in combination with paraoxon. The absolute NTE activity is calculated from the difference between the phenylvalerate concentrations in the leukocyte samples containing only paraoxon and in those samples that also contain mipafox.

2 Equipment, chemicals and solutions

2.1 Equipment

Single-beam or double-beam spectrometer or filter photometer with the possibility of measurement at 596 and 492 nm, including suitable cuvettes

Disposable syringes containing an anticoagulant (e.g. EDTA Vacutainer, Beckton-Dickinson)

LeucoSep® tubes for leukocyte separation (e.g. from Greiner Labortechnik)

Analytical balance (e.g. from Mettler)

Centrifuge (e.g. from Heraeus)

Shaking water bath, permitting thermostatic control at 37 °C

Pipettes with 100 µL and 1000 µL tips (e.g. from Eppendorf)

Glass Pasteur pipettes

50 mL Pipette

10 mL, 25 mL, 100 mL, 500 mL and 1000 mL Volumetric flasks (e.g. from Brand)

1000 mL Glass beaker

Screw-capped test-tubes (e.g. from Brandt)

14 mL Plastic tubes (e.g. from Sarstedt)

4 mL Glass vials with screw caps (e.g. from Zefa Laborservice)

pH meter with measurement electrode

2.2 Chemicals

4-Aminoantipyrine (e.g. Sigma, No. A4382)

Bradford reagent (e.g. Bio-rad, No. 500-0001)

BSA (bovine serum albumin) (e.g. Sigma, No. A7030)

Deionised water (e.g. produced by means of Millipore® technology)

Dimethylformamide (DMF) (e.g. Fluka, No. 40248)

Dodecylsulphate sodium salt (e.g. Merck, No. 1.12533.0250)

Ethylenediaminotetraacetic acid disodium salt (disodium EDTA) (e.g. Sigma, No. ED2SS)

Separation medium (for leukocyte separation, e.g. Ficoll-Paque®, Pharmacia, No. 17-0840-03)

Potassium hexacyanoferrate(III) (e.g. Fluka, No. 60299)

Sodium chloride solution, 0.9% (e.g. Braun Melsungen, No. 03730972)

Mipafox (from Bayer AG)

Paraoxon (diethyl-p-nitrophenyl phosphate) (e.g. Sigma D9286)

Phenol (e.g. Sigma P5566)

Phenylvalerate prepared according to reference [8]

Hydrochloric acid, 30% (e.g. Fluka 17077)

Tris base (e.g. Sigma T1503)

Triton X-100 (e.g. Sigma T9284)

2.3 Solutions

1 M Hydrochloric acid:
About 250 mL deionised water are placed in a 500 mL volumetric flask. Using a pipette 50 mL hydrochloric acid (30%) are added. After pipetting the concentrated hydrochloric acid, the pipette is rinsed several times with the contents of the flask to remove any residual hydrogen chloride gas from the body of the pipette. The flask is subsequently filled to its nominal volume with deionised water. The solution is kept at +4 to +6 °C in the refrigerator and can be used over a period of 6 months.

50 mM Tris EDTA buffer (pH 8):
6.055 g Tris base (50 mM) and 0.0745 g disodium EDTA (0.2 mM) are weighed exactly into a 1000 mL glass beaker and dissolved in approx. 800 mL deionised water. 1 M hydrochloric acid is added drop by drop using a pipette, while the pH is being checked constantly with a pH meter and the contents are being continuously stirred, until the desired pH value is reached. Then the contents are transferred to a 1000 mL volumetric flask and the flask is filled to its nominal volume with deionised water. The buffer solution is stable for 4 weeks when stored at +4 to +6 °C.

Stop/colour solution I (dodecylsulphate/4-aminoantipyrine solution):
1 g dodecylsulphate disodium salt and 0.025 g 4-aminoantipyrine are weighed exactly into a 100 mL volumetric flask and dissolved in Tris EDTA buffer. The volumetric flask is then filled to its nominal volume with Tris EDTA buffer. The solution must be freshly prepared before each analytical series.
Stop/colour solution I is added to denature the proteins (by means of dodecylsulphate) and thus to inactivate enzymes. It contains 4-aminoantipyrine, which reacts with phenol under oxidative conditions to form a quinonimine dye.

Colour solution II (0.4% potassium hexacyanoferrate(III) solution):
0.4 g potassium hexacyanoferrate(III) is weighed exactly into a 100 mL volumetric flask and dissolved with deionised water. The flask is subsequently filled to its nominal volume with deionised water. The solution must be freshly prepared before each analytical series.
Due to the oxidative properties of potassium hexacyanoferrate(III), colour solution II starts the reaction between phenol and 4-aminoantipyrine to form the quinonimine dye that can be detected at 492 nm by photometry.

0.03% Triton X-100 solution:
30 µL Triton X-100 are pipetted into a 100 mL volumetric flask. The flask is subsequently filled to its nominal volume with deionised water.
The Triton X-100 solution is stable at room temperature for at least 6 weeks.

Phenylvalerate stock solution (1.5%):
150 µL phenylvalerate are pipetted into a 10 mL volumetric flask. The flask is subsequently filled to its nominal volume with DMF. The solution is stable for up to 3 weeks when stored at +4 to +6 °C in the dark.

Substrate solution (0.06% phenylvalerate in 0.03% Triton X-100 solution):
1.0 mL of the 1.5% phenylvalerate solution is pipetted into a 25 mL volumetric flask. The flask is subsequently filled to its nominal volume with the 0.03% Triton X-100 solution. The solution must be freshly prepared before use in each case.

Paraoxon stock solution:
13.8 mg paraoxon are weighed exactly into a 25 mL volumetric flask and dissolved in Tris EDTA buffer. The volumetric flask is then filled to its nominal volume with Tris EDTA buffer (2 mmol/L).

Paraoxon working solution:
1.0 mL of the paraoxon stock solution is pipetted into a 10 mL volumetric flask. The volumetric flask is then filled to its nominal volume with Tris EDTA buffer (200 µmol/L).

Mipafox stock solution:
9.1 mg mipafox are weighed exactly into a 25 mL volumetric flask and dissolved in Tris EDTA buffer. The flask is subsequently filled to its nominal volume (2 mmol/L).

Mipafox working solution:
1.0 mL of the mipafox stock solution is pipetted into a 10 mL volumetric flask. The volumetric flask is then filled to its nominal volume with Tris EDTA buffer (200 µmol/L).

The paraoxon and mipafox stock and working solutions must be freshly prepared before each analytical series.

2.4 Calibration standards

2.4.1 Standard solutions for protein assay

BSA stock solution:
10 mg BSA (bovine serum albumin) are weighed into a 10 mL volumetric flask and dissolved with deionised water. The flask is subsequently filled to its nominal volume with deionised water (1 g/L).

BSA calibration standard solutions for protein assay:
The calibration standard solutions used to calibrate the protein assay are prepared by dilution of the BSA stock solution with deionised water in 10 mL volumetric flasks according to the pipetting scheme in Table 1.

Table 1. Pipetting scheme for the calibration standard solutions for protein assay

BSA calibration standard solution No.	Volume of the BSA stock solution [µL]	Final volume of the calibration standard solution [mL]	Concentration of BSA [µg/mL]
1	0	10	0
2	20	10	2
3	40	10	4
4	60	10	6
5	80	10	8
6	100	10	10

The BSA stock solutions and the calibration standard solutions must be freshly prepared before each analytical series.

2.4.2 Standard solutions to determine the activity of NTE

Phenol stock solution:
40 mg phenol are weighed exactly into a 10 mL volumetric flask and dissolved in DMF. The flask is subsequently filled to its nominal volume (4 g/L). The solution must be freshly prepared before use in each case.

Phenol working solution:
Initially eight phenol working solutions are prepared by dilution of the phenol stock solution with DMF in eight 4 mL glass vials. The pipetting scheme for the preparation is shown in Table 2. These solutions must be freshly prepared before each analytical series.

Table 2. Pipetting scheme for phenol working solutions

Phenol working solution No.	Volume of DMF [μL]	Volume of phenol stock solution [μL]	Concentration of phenol [mg/L]
1	2000	0	0
2	1980	20	40
3	1950	50	100
4	1900	100	200
5	1800	200	400
6	1700	300	600
7	1600	400	800
8	1500	500	1000

Phenol calibration standard solutions to determine the NTE activity:
The calibration standard solutions are prepared by dilution of phenol working solutions 1 to 8 with Triton X-100 solution.
1 mL of each working solution is pipetted into a 25 mL volumetric flask. The flask is subsequently filled to its nominal volume with Triton X-100 solution. The phenol concentrations of calibration standard solutions 1 to 8 are shown in Table 3.

Table 3. Phenol calibration standard solutions to determine the NTE activity

Phenol calibration standard solution No.	Concentration of phenol [μg/2 mL]	Corresponding concentration of phenylvalerate* [nM/2 mL]
1	0	0
2	3.2	34.0
3	8.0	85.01
4	16.0	170.01
5	32.0	340.03
6	48.0	510.04
7	64.0	680.06
8	80.0	850.07

* The concentration of phenylvalerate that would give rise to the phenol concentrations in the middle column after complete enzymatic cleavage.

The working solutions and the calibration standard solutions to determine the activity of NTE must be freshly prepared on each working day, as otherwise concentration losses are to be expected.

3 Specimen collection and sample preparation

The blood withdrawn in EDTA Vacutainers (a total of 10 mL) is further processed immediately or stored for up to 24 hours at +4 to +6 °C.

3.1 Sample preparation

The leukocytes are isolated from 5 mL blood with the aid of LeucoSep® tubes as follows: 3 mL of the separation medium are pipetted into LeucoSep® tubes, then the contents are centrifuged for 30 s at 1000 g and room temperature. Subsequently 5 mL stabilised whole blood are added with a pipette and the mixture is centrifuged for 20 min at 800 g. Four layers are then present in the tube: plasma – leukocytes – separation medium – erythrocytes. A Pasteur pipette is used to withdraw the entire white leukocyte ring without sucking in the other phases. The leukocyte solution is transferred to a plastic tube and centrifuged for 5 minutes at 1200 g, the supernatant is discarded. Then 0.5 to 1 mL of the 0.9% sodium chloride solution is added to the leukocytes, the contents are mixed and centrifuged again for 5 minutes at 1200 g, the supernatant is discarded. The pellet is carefully resuspended in 1 mL of the 0.9% sodium chloride solution. Aliquots of this suspension are taken for the protein assay and also for the determination of the NTE activity.

4 Operational parameters

The spectrometer is adjusted to a wavelength of 596 nm (for the protein assay) or 492 nm (for the NTE activity determination), or an appropriate filter is used.

5 Analytical determination

5.1 Protein assay

For the protein assay two 20 µL aliquots of the suspension prepared as described in Section 3.1 are each pipetted into a 14 mL plastic tube. Each tube is filled to a volume of 2 mL with deionised water and allowed to stand for approx. 5 minutes to ensure that the cells have burst. Two volumes of 800 µL of this leukocyte haemolysate are each pipetted into 14 mL plastic tubes (2 duplicate determinations per sample) and the following pipetting scheme is followed (Table 4).
The protein assay is carried out photometrically by means of the Bradford test. This is based on the binding of the ionic dye Coomassie Brilliant Blue G 250 to basic amino acids, which causes a shift in the absorption maximum of the dye from 465 nm to 596 nm. The absorption at 596 nm is recorded by the spectrometer with respect to the blank, and the protein concentration is determined on the basis of an ex-

Table 4. Pipetting scheme for protein assay

	Water	Leukocytes	BSA calibration standard solution 1 to 6	Bradford reagent
Blank value	800 µL			200 µL
Calibration curve			800 µL	200 µL
Sample		800 µL		200 µL

ternal linear calibration graph (see Figure 2). It is obtained by plotting the measured absorption of BSA calibration standard solutions 1 to 6 as a function of the concentrations used. The measured absorption of the sample is used to read off the appropriate protein concentration in µg per litre from the relevant calibration graph. The mean of the values obtained for the multiple protein assay (a total of 4 per sample) is calculated.

5.2 NTE activity determination

50 µL of the suspension prepared as described in Section 3.1 are pipetted into a 14 mL plastic tube. The tube is filled to a volume of 5 mL with deionised water and allowed to stand for approx. 5 minutes to ensure that the cells have burst. A concentration of approx. 200 µg protein/mL is required for NTE determination. Therefore the leukocyte haemolysate must be appropriately diluted with Tris EDTA buffer before the determination.

In each case 1 mL of the adjusted leukocyte haemolysate is filled into 4 sealable test-tubes with a capacity of approx. 10 mL (2 sample blanks and 2 samples) and they are treated as shown in the pipetting scheme shown in Table 5. Final photometric determination is carried out at 492 nm with respect to deionised water as a blank. The absorption of the samples is determined in duplicate.

6 Calibration

Calibration is carried out using pooled leukocyte haemolysate with its protein content adjusted to 200 µg/mL. The phenol calibration standard solutions prepared as described in Section 2.4.2 are treated according to the pipetting scheme (Table 6), mixed thoroughly and then incubated for 15 minutes at room temperature. The samples are subsequently measured at 492 nm in the spectrometer with respect to deionised water as a blank.

The calibration functions are obtained by plotting the measured absorption as a function of the concentrations used. The gradient of the calibration function and the intercept with the y-axis are calculated by linear regression.

Table 5. Pipetting scheme for the samples for determination of NTE activity

	Adjusted leukocyte haemolysate [mL]	Tris EDTA buffer [mL]	Paraoxon working solution [mL]	Mipafox working solution [mL]	Incubation	Stop/ colour solution I [mL]	Substrate solution [mL]	Incubation	Stop/ colour solution I [mL]	Colour solution II [mL]	Incubation and measurement
Sample blank + paraoxon	1.0	0.5	0.5	–	20 min at 37 °C in shaking bath	2.0	2.0	60 min at 37 °C in shaking bath	–	1.0	Mix well, leave to stand for 15 min at RT, measure by photometry at 492 nm with respect to water
Sample + paraoxon	1.0	0.5	0.5	–		–	2.0		2.0	1.0	
Sample blank + paraoxon + mipafox	1.0	–	0.5	0.5		2.0	2.0		–	1.0	
Sample + paraoxon + mipafox	1.0	–	0.5	0.5		–	2.0		2.0	1.0	

Table 6. Pipetting scheme for calibration of the determination of NTE activity

Tris EDTA buffer [mL]	Leukocyte haemolysate [mL]	Stop/colour solution I [mL]	Phenol calibration standard solution 1 to 8 [mL]	Colour solution II [mL]
1.0	1.0	2.0	2.0	1.0

The calibration function is linear over the entire concentration range (0 to 850 nmol phenol/2 mL). Figure 3 shows an example of the linear calibration function for determination of the NTE activity.

7 Calculation of the analytical result

The mean value of the sample absorptions obtained in a duplicate determination as described in Section 5.2 is calculated and the mean value for the sample blank is subtracted from the result.

The phenylvalerate concentration is calculated on the basis of an external linear calibration function (see Section 6). Taking into account the absorptions for the leukocyte haemolysate spiked with paraoxon only or with mipafox and paraoxon (according to Section 5.2, Table 5) that have been and corrected by the amount of the sample blank, the pertinent phenylvalerate concentration in nmol can be read off the relevant calibration graph.

The absolute NTE activity is calculated from the following equation:

$$\text{NTE activity} = \frac{c\,(P - PM)}{60 \times \text{conc. protein}} \; \frac{\text{nmol}}{\min \times \text{mg}}$$

c (P–PM) = Difference between the phenylvalerate concentration of the sample spiked with paraoxon and the phenylvalerate concentration of the sample spiked with mipafox + paraoxon

8 Evaluation of the method

8.1 Precision

The leukocytes were isolated from a blood sample and after dilution with Tris EDTA buffer a protein assay was carried out in order to determine the precision in the series with respect to the protein content. Determination was performed 20 times thus resulting in a relative standard deviation of 10.6% (at a mean protein content of 40 µg/mL buffer).

Leukocyte haemolysate was spiked with a defined quantity of phenol (40 mg/L) in order to ascertain the precision of the phenol determination. Then the phenol concentration was determined 10 times (see Table 7).
The precision of the determination of the NTE activity was not determined.

Table 7. Precision in the series for the determination of protein content and for phenol in leukocyte haemolysate

	n	Standard deviation (rel.) [%]	Prognostic range [%]
Protein content	20	10.6	22.2
Phenol determination	10	1.8	3.9

It proved impossible to ascertain the precision of the NTE activity determination from day to day, as the activity of NTE in a blood sample remains stable for at most 24 h.

8.2 Accuracy

It was not possible to determine the accuracy, as no suitable reference standards were available.

8.3 Detection limit

A detection limit for the determination of the NTE activity in the normal sense (e.g. as three times the signal/background noise ratio in the temporal vicinity of the analytical signal) cannot be given, as it is masked by the natural NTE activity of the native leukocytes.

8.4 Sources of error

The method must be performed in a reproducible manner. As in the case of all enzymatic determinations, it is important to maintain a constant temperature, 37 °C in this case.
The proposal to base the result on the protein content may lead to variations between laboratories, and therefore trained personnel should always perform the protein assay.

9 Discussion of the method

The NTE activity in human leukocytes is determined using this method.
NTE activities below 3.0 nmol phenylvalerate/min/mg protein may be regarded as non-physiological, as they were not observable in practice [3].
As an alternative to the protein assay, the NTE activity can be based on the leukocyte count [5]. Mipafox and phenylvalerate are no longer commercially available. Mipafox can be obtained from Bayer for the time being. The procedure for the preparation of phenylvalerate is comprehensively described in the literature [8].

Instruments used:
HP 8453 UV-Vis diode array spectrometer from Hewlett-Packard

10 References

[1] *M. B. Abou Donia* and *D. M. Lapadula:* Mechanisms of organophosphorus ester-induced delayed neurotoxicity: Type I and Type II. Annu. Rev. Pharmacol. Toxicol. 30, 405–440 (1990).
[2] *R. J. Richardson:* Interactions of organophosphate compounds with neurotoxic esters. In: *J. E. Chambers* and *P. E. Levi* (eds.): Organophosphates: Chemistry, Fate, and Effects. Academic press. Inc. (1992).
[3] *P. Ruhnau, M. Müller, E. Hallier* and *J. Lewalter:* Ein validiertes photometrisches Verfahren für die Bestimmung der Neuropathy Target Esterase Aktivität in humanen Lymphocyten. Umweltmed. Forsch. Prax. 6(2), 101–103 (2001).
[4] *D. Bertoncin, A. Russolo, S. Caroldi* and *M. Lotti:* Neuropathy target esterase in human lymphocytes. Arch. Env. Health 40(3), 139–143 (1985).
[5] *M. Lotti, S, Carold* and *A. Moretto:* Blood copper in organophosphate-induced delayed polyneuropathy. Tox. Letters 41, 175–180 (1988).
[6] *Deutsche Forschungsgemeinschaft:* List of MAK and BAT Values 2006, 42nd report, Wiley-VCH, Weinheim (2006).
[7] *H. Greim* and *G. Lehnert* (eds.): Acetylcholine Esterase Inhibitors. Biological Exposure Values for Occupational Toxicants and Carcinogens – Critical Data Evaluation for BAT and EKA Values. Volume 2. Wiley-VCH, Weinheim (1995).
[8] *M. K. Johnson:* Improved assay of neurotoxic esterase for screening organophosphates for delayed neurotoxicity potential. Arch. Toxicol. 37, 113–115 (1977).

Authors: *J. Lewalter, G. Leng, B. Röhrig*
Examiner: *M. Müller*

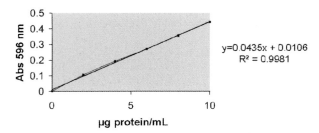

Fig. 2. Protein linear calibration graph

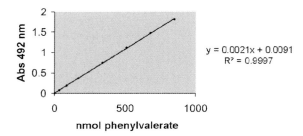

Fig. 3. Phenol linear calibration graph for NTE determination

Organophosphates (chlorpyrifos, diazinon, fenitrothion, fenthion, malathion)

Application   Determination in whole blood

Analytical principle   Gas chromatography/mass selective detection

Summary

The procedure described here permits the assay of the organophosphate insecticides chlorpyrifos, diazinon, fenitrothion, fenthion and malathion in whole blood. Exposure to these organophosphates that is of relevance to occupational medicine can be determined.

For this purpose whole blood stabilised with EDTA undergoes haemolysis in an ultrasonic bath. The organophosphates therein are enriched and purified on kieselguhr cartridges (Extrelut NT 20). After elution, the analytes are concentrated in a stream of nitrogen. On completion of separation by capillary gas chromatography, the samples are measured by means of a mass selective detector and electron impact ionisation (EI).

Calibration curves for quantitative evaluation are plotted using calibration standards prepared with pooled human blood. The calibration standards are treated in the same manner as the whole blood samples to be investigated. Chlorpyrifos-diethyl-D_{10} is added to the blood samples as an internal standard. The detection limit of the method is between 0.5 and 2.6 µg per litre blood for the individual organophosphates.

Diazinon

Within-series imprecision: Standard deviation (rel.) s_w = 3.7%, 3.4% or 11.9%
                            Prognostic range u = 9.5%, 8.7% or 30.6%
                            at a spiked concentration of 8, 16 or 20 µg
                            per litre whole blood and where n = 6 determinations

The MAK-Collection Part IV: Biomonitoring Methods, Vol. 11.
DFG, Deutsche Forschungsgemeinschaft
Copyright © 2008 WILEY-VCH Verlag GmbH & Co. KGaA, Weinheim
ISBN: 978-3-527-31596-3

Organophosphates

Between-day imprecision: Standard deviation (rel.) $s_w = 3.0\%$
Prognostic range $u = 7.7\%$
at a spiked concentration of 16 µg per litre
whole blood and where n = 6 determinations

Accuracy: Recovery rate $r = 100\%$ at 8 µg/L
and 91% at 20 µg/L

Detection limit: 2.3 µg per litre whole blood

Fenitrothion

Within-series imprecision: Standard deviation (rel.) $s_w = 8.0\%$, 10.1% or 9.9%
Prognostic range $u = 20.6\%$, 26% or 25.4%
at a spiked concentration of 8, 16 or 20 µg
per litre whole blood and where n = 6 determinations

Between-day imprecision: Standard deviation (rel.) $s_w = 12.9\%$
Prognostic range $u = 33.2\%$
at a spiked concentration of 16 µg per litre
whole blood and where n = 6 determinations

Accuracy: Recovery rate $r = 120\%$ at 8 µg/L
and 129% at 20 µg/L

Detection limit: 2.6 µg per litre whole blood

Malathion

Within-series imprecision: Standard deviation (rel.) $s_w = 12.8\%$, 9.2% or 8.8%
Prognostic range $u = 32.9\%$, 23.6% or 22.6%
at a spiked concentration of 8, 16 or 20 µg
per litre whole blood and where n = 6 determinations

Between-day imprecision: Standard deviation (rel.) $s_w = 10.6\%$
Prognostic range $u = 27.2\%$
at a spiked concentration of 16 µg per litre
whole blood and where n = 6 determinations

Accuracy: Recovery rate $r = 92\%$ at 8 µg/L
and 118% at 20 µg/L

Detection limit: 2.2 µg per litre whole blood

Fenthion

Within-series imprecision: Standard deviation (rel.) s_w=3.6%, 3.4% or 5.3%
Prognostic range u=9.3%, 8.7% or 13.6%
at a spiked concentration of 8, 16 or 20 µg
per litre whole blood and where n=6 determinations

Between-day imprecision: Standard deviation (rel.) s_w=2.6%
Prognostic range u=6.7%
at a spiked concentration of 16 µg per litre
whole blood and where n=6 determinations

Accuracy: Recovery rate r=108% at 8 µg/L
and 106% at 20 µg/L

Detection limit: 0.5 µg per litre whole blood

Chlorpyrifos

Within-series imprecision: Standard deviation (rel.) s_w=6.6%, 5.3% or 4.4%
Prognostic range u=17%, 13.6% or 11.3%
at a spiked concentration of 8, 16 or 20 µg
per litre whole blood and where n=6 determinations

Between-day imprecision: Standard deviation (rel.) s_w=4.9%
Prognostic range u=12.6%
at a spiked concentration of 16 µg per litre
whole blood and where n=6 determinations

Accuracy: Recovery rate r=103% at 8 µg/L
and 122% at 20 µg/L

Detection limit: 2.5 µg per litre whole blood

Organophosphates

The substances that can be determined by this method belong to the organophosphate group of compounds (organic esters of phosphoric acid). Based on their chemical structure, chlorpyrifos, diazinon, fenitrothion and fenthion are assigned to the triesters of thiophosphoric acid, and malathion to the triesters of dithiophosphoric acid (cf. Fig. 1). They are used as active substances in plant protection agents [1] and disinfestation agents [2] on account of their insecticidal and partly also acaricidal properties.

The above-mentioned substances have a very selective effect in the control of arthropods. However, they also represent a source of risk to humans due to their high tox-

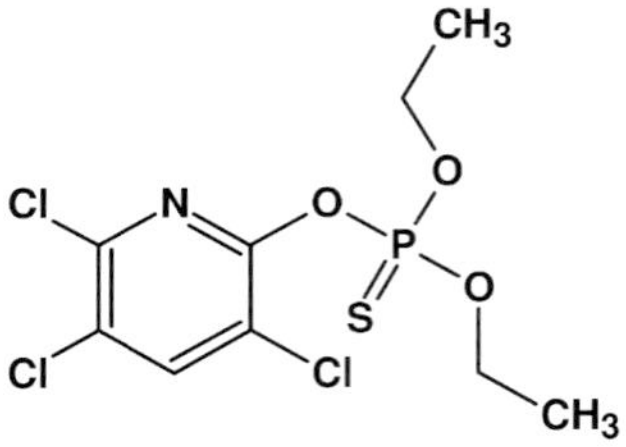

Chlorpyrifos

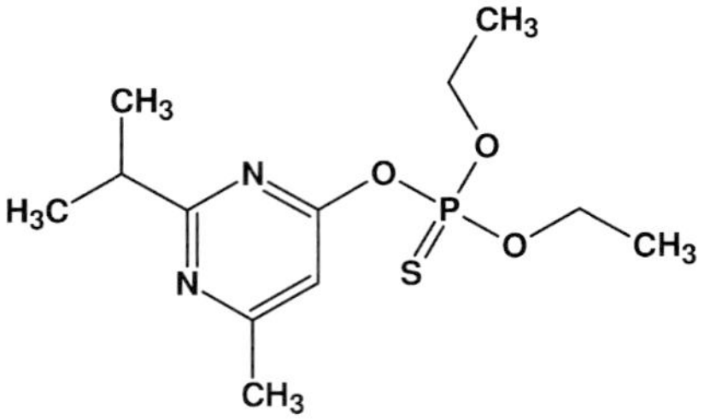

Diazinon

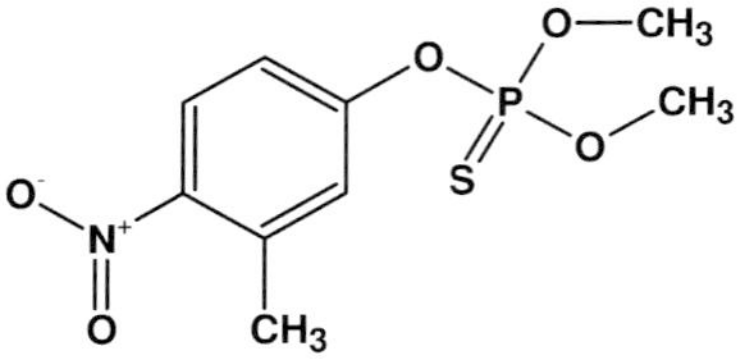

Fenitrothion

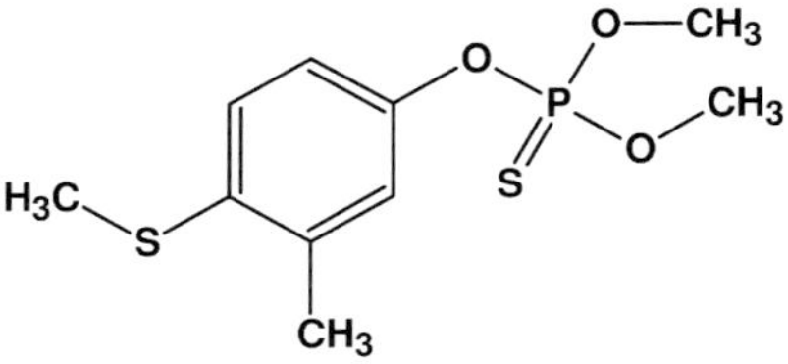

Fenthion

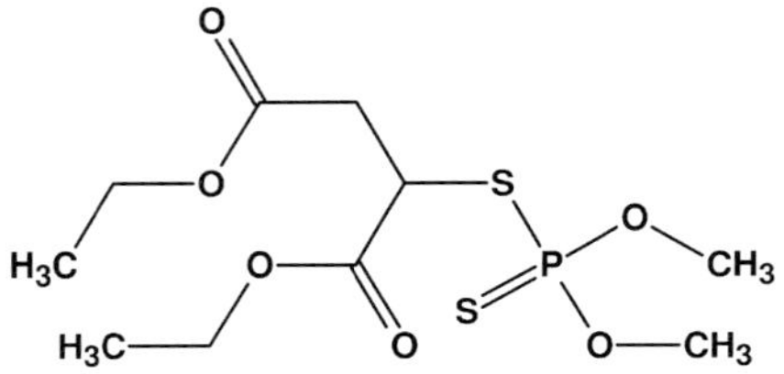

Malathion

Fig. 1. Structural formulae of the organophosphates that are detectable using this method

Table 1. MAK and ADI values

Substance	CAS number	ADI [3] [mg/kg body weight per day]	Classification by the Commission	
			MAK values [4] [mg/m^3]	Documentation
Chlorpyrifos	2921-88-2	0.01		
Diazinon	333-41-5	0.002	0.1	[5]
Fenitrothion	122-14-5	0.005		
Fenthion	55-38-9	0.007	0.2	[6]
Malathion	121-75-5	0.3	15	[7]

icity. This applies to inappropriate use in confined spaces and in particular to occupational handling of these agents if the safety requirements and instructions for use are not properly observed (industrial manufacturing, formulation, filling, performance of pest control measures).

The MAK values (maximum permissible concentration at the workplace) and the ADI values (acceptable daily intake) derived from the toxicological data are shown in Table 1.

Authors: *E. Berger-Preiss, S. Gerling*
Examiners: *J. Lewalter, K. Wilmersdorf*

Organophosphates (chlorpyrifos, diazinon, fenitrothion, fenthion, malathion)

Application Determination in whole blood

Analytical principle Gas chromatography/mass selective detection

Contents

1 General principles
2 Equipment, chemicals and solutions
2.1 Equipment
2.2 Chemicals
2.3 Solutions and conditioning of the kieselguhr cartridges
2.4 Calibration standards
3 Specimen collection and sample preparation
3.1 Sample preparation
4 Operational parameters
4.1 Operational parameters for gas chromatography and mass spectrometry
5 Analytical determination
6 Calibration
7 Calculation of the analytical result
8 Standardisation and quality control
9 Evaluation of the method
9.1 Precision
9.2 Accuracy
9.3 Detection limits
9.4 Sources of error
9.4.1 Special aspects to be observed during sample processing and analysis
9.4.2 Investigations on the stability of the organophosphates in whole blood
10 Discussion of the method
11 References

The MAK-Collection Part IV: Biomonitoring Methods, Vol. 11.
DFG, Deutsche Forschungsgemeinschaft
Copyright © 2008 WILEY-VCH Verlag GmbH & Co. KGaA, Weinheim
ISBN: 978-3-527-31596-3

1 General principles

Whole blood stabilised with EDTA undergoes haemolysis in an ultrasonic bath. The organophosphates therein are enriched and purified on kieselguhr cartridges (Extrelut NT 20). After elution, the analytes are concentrated in a stream of nitrogen. On completion of separation by capillary gas chromatography, the samples are measured by means of a mass selective detector and electron impact ionisation (EI).

Calibration curves are plotted for quantitative evaluation using calibration standards prepared with pooled human blood. The calibration standards are treated in the same manner as the whole blood samples to be investigated. Chlorpyrifos-diethyl-D_{10} is added to the blood samples as an internal standard. The detection limit of the method is between 0.5 and 2.6 µg per litre blood for the individual organophosphates.

2 Equipment, chemicals and solutions

2.1 Equipment

Gas chromatograph with split/splitless injector, mass selective detector, autosampler and data processing system

Capillary gas chromatographic column:
Length: 60 m; inner diameter: 0.25 mm; stationary phase:
(5% phenyl)methylpolysiloxane; film thickness: 0.25 µm (e.g. HP 5-MS from Agilent)

Disposable syringes containing an anticoagulant (e.g. K_2-EDTA Kabevettes from Kabe)

Ultrasonic bath

Laboratory centrifuge

Vacuum station for solid phase extraction (e.g. from Supelco)

PTFE valves for the vacuum station

Water jet pump or vacuum pump

Vortex mixer

Device for evaporation under a stream of nitrogen

25 to 50 mL Centrifuge vials with screw cap and PTFE coated inner septum

25 to 50 mL Centrifuge vials with conical lower part and graduation

Pasteur pipettes

10 mL Volumetric flasks

25 mL, 100 mL, 500 mL, and 1000 mL Measuring cylinders

Organophosphates

1000 mL Glass beaker

1 L Amber glass bottle

1.8 mL Crimp-top autosampler vials with 250 µL microvial inserts

Microlitre pipettes, adjustable between 1 and 10 µL, 10 and 100 µL, and 100 and 1000 µL (e.g. from Eppendorf)

10 µL Syringe for gas chromatography

2.2 Chemicals

Extrelut® NT20 cartridges (Merck No. 1.15096.0001)

Deionised water (e.g. produced by means of Millipore® technology)

Acetone (e.g. Sigma Aldrich, Pestanal, No. 34480)

Ethyl acetate (e.g. Sigma Aldrich, Pestanal, No. 34490)

Dichloromethane SupraSolv® (e.g. Merck, No. 106054)

Toluene SupraSolv® (e.g. Merck, No. 108389)

n-Hexane (e.g. from Rathburn)

Chlorpyrifos (e.g. Riedel-de Haën, Pestanal, No. 45395)

Chlorpyrifos-diethyl-D_{10} (e.g. Sigma-Aldrich, No. 488569)

Diazinon (e.g. Sigma Aldrich, Pestanal, No. 45428)

Fenitrothion (e.g. Sigma Aldrich, Pestanal, No. 45487)

Fenthion (e.g. Sigma Aldrich, Pestanal, No. 36552)

Malathion (e.g. Sigma Aldrich, Pestanal, No. 36143)

Bromophos-methyl (e.g. Sigma Aldrich, Pestanal, No. 33400)

Helium 5.0 (e.g. from Linde)

2.3 Solutions and conditioning of the kieselguhr cartridges

Ethyl acetate/dichloromethane (2+1) elution solution:
600 mL ethyl acetate and 300 mL dichloromethane, both of trace analysis grade (Pestanal), are mixed in a 1000 mL glass beaker and then stored in a 1 L amber glass bottle. The elution solution should be kept in a sealed vessel at 4 to 6 °C and is stable for at least 3 months under these conditions.

Solution of the internal standard

Starting solution of the internal standard:
10 mg chlorpyrifos-diethyl-D_{10} (ISTD) are weighed exactly in a 10 mL volumetric flask and dissolved with acetone, then the flask is filled to its nominal volume. This solution is stable for about 1 month when stored in the dark at 4 to 6 °C. The concentration of this solution is 1 g/L.

Spiking solution of the internal standard:
100 µL of the starting solution of the internal standard are pipetted into a 10 mL volumetric flask and the flask is filled up to its nominal volume with acetone. The concentration of this solution is 10 mg/L.

Chromatography standard

Starting solution of the chromatography standard:
10 mg bromophos-methyl are weighed exactly in a 10 mL volumetric flask and dissolved with toluene, then the flask is filled to its nominal volume with toluene. This solution is stable for at least 3 months when stored in the dark at 4 to 6 °C. The concentration of this solution is 1 g/L.

Spiking solution of the chromatography standard:
100 µL of the starting solution of the chromatography standard are pipetted into a 10 mL volumetric flask, and the flask is filled up to its nominal volume with toluene. The concentration of this solution is 10 mg/L.

The spiking solutions of the internal standard and the chromatography standard are stored in the dark in sealed vessels at 4 to 6 °C, and they are stable under these conditions for at least 1 month.

Conditioning of the Extrelut NT-20 cartridge

The cartridges are placed on the vacuum station for solid phase extraction. The PTFE valve is opened and 20 mL deionised water are introduced into the cartridge. Permeation of the water into the kieselguhr can be accelerated by applying a slight vacuum (the kieselguhr should be completely moist). The water is allowed to act for 30 minutes while the valve is closed. After opening the PTFE valve, 30 mL of the elution solution (ethyl acetate/dichloromethane (2+1)) are introduced into the cartridge, whereby the solvent is allowed to drip slowly through the cartridge. Then the cartridge is conditioned in the same manner with 30 mL hexane, followed by 50 mL acetone. A slight vacuum is applied to dry the cartridge. The cartridge must be dry before the sample is introduced, as otherwise problems may occur.

2.4 Calibration standards

Starting solutions:
Approx. 10 mg of each of the organophosphates chlorpyrifos, diazinon, fenitrothion, fenthion and malathion are weighed exactly in separate 10 mL volumetric flasks and dissolved with acetone, then the flasks are filled to the nominal volume with acetone. These solutions are stable in the refrigerator for at least 1 month. The concentration of these solutions is 1 g/L.

Stock solution:
1000 µL of each starting solution are pipetted into a 10 mL volumetric flask, and the flask is filled to its nominal volume with acetone. The concentration of this solution is 100 mg/L.

Spiking solution A:
1000 µL of the stock solution are pipetted into a 10 mL volumetric flask, and the flask is filled to its nominal volume with acetone. The concentration of this solution is 10 mg/L.

Spiking solution B:
100 µL of the stock solution are pipetted into a 10 mL volumetric flask and the flask is filled to its nominal volume with acetone. The concentration of this solution is 1 mg/L.

The starting and spiking solutions are stored in sealed vessels at 4 to 6 °C, and they are stable under these conditions for at least 1 month.

The standard solutions used for calibration are prepared in a centrifuge vial (25 to 50 mL) with a screw cap by spiking 5 mL pooled, haemolysed human blood in each case. The pipetting scheme for the preparation is shown in Table 2. The error resulting from dilution is negligible. These calibration standard solutions must be freshly prepared before each analytical series.

Table 2. Pipetting scheme for the preparation of the calibration standard solutions

Volume of the spiking solutions		Volume of the spiking solution of the ISTD	Volume of the pooled human blood	Concentration of the calibration standards
A [µL]	B [µL]	[µL]	[mL]	[µg/L]
–	20	25	5	4
4	–	25	5	8
8	–	25	5	16
12	–	25	5	24
16	–	25	5	32

3 Specimen collection and sample preparation

K_2-EDTA Kabevettes are used to withdraw blood. The blood samples should be processed as soon as possible, but at the latest after 1 day of storage at 4 to 6 °C in the refrigerator. A blood sample of at least 6 mL should be taken.

3.1 Sample preparation

5 mL K_2-EDTA whole blood are pipetted into a screw-capped centrifuge vial (25 to 50 mL) and the sample is treated in the ultrasonic bath for 15 minutes in order to ensure haemolysis of the erythrocytes. Then 25 µL of the spiking solution of the internal standard are added. Finally, 7 mL acetone are carefully added, whereby mixing with the blood must be avoided.

After this addition, the solution is mixed vigorously for approx. 20 s on the vortex mixer. A granular protein precipitate is formed. Then the sample is centrifuged (3000 rpm, 30 s). After centrifugation, a Pasteur pipette is used to transfer the supernatant liquid to the Extrelut NT-20 cartridge conditioned as described in Section 2.3 (with the PTFE valve open) and it is allowed to seep slowly into the cartridge at atmospheric pressure. After the sample has seeped into the cartridge, it is allowed to stand for 30 minutes with the valve closed.

During this time 20 mL of the elution solution (ethyl acetate/dichloromethane (2 + 1)) are added to the centrifugation residue, the mixture is swirled briefly and then treated for 5 minutes in the ultrasonic bath. After another centrifugation step (3000 rpm, 30 s) the supernatant liquid is introduced onto the cartridge. A vessel (e.g. a centrifuge vial with a conical lower part and graduation from 25 to 50 mL) is placed under the cartridge to collect the eluate. Now the valve is opened and the flow rate of the eluate must be set at approx. 1 drop per second.

At the same time 20 mL n-hexane are added to the remaining centrifugation residue, which is then swirled briefly, treated for 5 minutes in the ultrasonic bath and centrifuged anew (3000 rpm, 30 s). Then the n-hexane residue is transferred to the same cartridge and the eluate is collected in the collection vessel. The volume of the eluate is about 20 mL, as a large part of the elution mixture remains in the cartridge. If necessary, a slight vacuum should be applied.

200 µL toluene are added to the eluate as a keeper. The eluate is subsequently evaporated to about 150 µL in a stream of nitrogen. If two phases are formed during the evaporation step, the upper yellow phase is withdrawn using a pipette and processing is continued. After adding 25 µL of the spiking solution of the chromatography standard (bromophos-methyl) the solution is filled to a volume of approx. 200 µL with toluene. Finally, the solution is transferred to an autosampler glass insert and sealed tightly.

4 Operational parameters

4.1 Operational parameters for gas chromatography and mass spectrometry

Capillary column:	Material:	Fused silica
	Stationary phase:	HP 5-MS
	Length:	60 m
	Inner diameter:	0.25 mm
	Film thickness:	0.25 µm
Detector:	Mass selective detector (MSD)	
Temperatures:	Column:	2 minutes at 60 °C; then increase by 30 °C/minute until 150 °C; then increase by 3 °C/minute until 280 °C; 15 minutes isothermal, then 10 °C/minute until 300 °C, 20 minutes at the final temperature
	Injector:	250 °C
	Transfer line:	280 °C
Carrier gas:	Helium 5.0 at a constant flow rate of 1.4 mL per minute	
Split:	Splitless, split on after 1 minute	
Sample volume:	1 µL	
Ionisation type:	Electron impact ionisation (EI)	
Ionisation energy:	70 eV	
Ion source temperature:	230 °C	
Quadrupole temperature:	150 °C	
Electron multiplier:	(Autotuning optimised)	

All other parameters must be optimised in accordance with the manufacturer's instructions.

5 Analytical determination

The operational parameters for the instruments are adjusted as stipulated and 1 µL of the analytical sample is injected into the gas chromatograph in each case. Figure 2 shows the chromatogram of a standard in toluene. Figure 3 shows an example of the GC/MS chromatogram of a whole blood sample spiked with 24 µg/L of the analyte.

A quality control sample and a reagent blank sample, consisting of 5 mL bidistilled water, are prepared as described in Section 3.1 and analysed in each analytical series. Quantification of the individual organophosphates is based on the masses shown in Table 3.

Table 3. Retention times and recorded masses

Analyte	Retention time [minutes]	Dwell time [ms]	Recorded mass [m/z]
Diazinon	30.22	100	137* 179
Fenitrothion	35.57	80	277* 125
Malathion	36.06	80	173* 125
Chlorpyrifos-diethyl-D_{10} (ISTD)	36.74	80	324*
Fenthion	36.87	80	278* 125
Chlorpyrifos	37.01	80	197* 314
Bromophos-methyl (chromatography standard)	38.38	80	331*

The masses marked * are used for quantitative evaluation.
ISTD = internal standard.

6 Calibration

The calibration standard solutions prepared as described in Section 2.4 are processed in the same manner as the samples (Section 3.1) and analysed by gas chromatography/mass spectrometry as stipulated in Sections 4 and 5. Two calibration standard solutions are processed to assure reliable results. Calibration curves are obtained by plotting the quotients of the peak areas for each organophosphate and of the internal standard as a function of the concentrations used. The gradient of the calibration curve and the intercept with the y-axis are calculated by linear regression.

The calibration functions of the organophosphates are each linear in the concentration range between 4 and 32 µg/L blood. Figure 4 shows the calibration curves of the individual organophosphates.

7 Calculation of the analytical result

The organophosphate concentration in whole blood samples is calculated on the basis of a linear calibration function (cf. Section 6). Quotients are calculated by dividing the peak areas of the analytes by that of the internal standard. These quotients are used to read off the pertinent concentration of the organophosphate in µg per litre blood from the relevant calibration graph.

If the whole blood used to prepare the calibration standard solutions exhibits background interference from organophosphates, the resulting calibration graph must be shifted in parallel so that it passes through the zero point of the coordinates. (The concentrations of the background exposure can be read off from the point where the graph intercepts the axis before parallel shifting in each case.) No such background interference has occurred in practice to date. Any reagent blank values that are found must be subtracted from the analytical results for the real samples.

8 Standardisation and quality control

Quality control of the analytical results is carried out as stipulated in the guidelines of the Bundesärztekammer (German Medical Association) [8] and in the special preliminary remarks to this series. A control sample containing a constant concentration of the individual organophosphates is analysed in order to check the precision of the method. As material for quality control is not commercially available, it must be prepared in the laboratory.

For this purpose, pooled human blood is spiked with a defined quantity of the individual organophosphates. Spiking should be carried out just before processing on account of the short period that the samples can be stored (cf. Section 3.1).

The expected value and the tolerance range of this quality control material are ascertained in a pre-analytical period (one analysis of the control material on each of 20 different days) [9, 10].

9 Evaluation of the method

9.1 Precision

Pooled human blood is spiked with a defined quantity of organophosphates, processed and analysed to determine the precision in the series. Six replicate assays of the organophosphates yielded the precision in the series documented in Table 4.

In addition, the precision from day to day was determined. For this purpose pooled human blood was spiked, processed and analysed on 6 different days. The resulting precision from day to day is also given in Table 5.

9.2 Accuracy

Recovery experiments were carried out at two concentrations to test the accuracy of the method. The same material was used as for the determination of the precision in the series. These blood samples were each processed 6 times as described in Section

Table 4. Precision in the series for the assay of organophosphates in whole blood (n=6)

Analyte	Concentration [µg/L blood]	Standard deviation (rel.) [%]	Prognostic range [%]
Diazinon	8	3.7	9.5
	16	3.4	8.7
	20	11.9	30.6
Fenitrothion	8	8.0	20.6
	16	10.1	26.0
	20	9.9	25.4
Malathion	8	12.8	32.9
	16	9.2	23.6
	20	8.8	22.6
Fenthion	8	3.6	9.3
	16	3.4	8.7
	20	5.3	13.6
Chlorpyrifos	8	6.6	17.0
	16	5.3	13.6
	20	4.4	11.3

Table 5. Precision from day to day for the determination of organophosphates in whole blood (n=6)

Analyte	Concentration [µg/L]	Standard deviation (rel.) [%]	Prognostic range [%]
Diazinon	16	3.0	7.7
Fenitrothion	16	12.9	33.2
Malathion	16	10.6	27.2
Fenthion	16	2.6	6.7
Chlorpyrifos	16	4.9	12.6

Table 6. Mean relative recovery rates for the organophosphates in whole blood samples

Analyte	n	Spiked concentration [µg/L]	Mean relative recovery [%]	Range [%]
Diazinon	6	8	100	95–103
	6	20	91	73–95
Fenitrothion	6	8	120	115–130
	6	20	129	115–134
Malathion	6	8	92	73–108
	6	20	118	105–132
Fenthion	6	8	108	100–110
	6	20	106	96–112
Chlorpyrifos	6	8	103	95–113
	6	20	122	113–128

Organophosphates

3 and analysed as stipulated in Sections 4 and 5. The mean relative recovery rates are shown in Table 6.

9.3 Detection limits

The detection limits for the individual analytes were calculated as three times the signal/background noise ratio in the temporal vicinity of the analyte signal. The detection limits for the parameters determined using this method are shown in Table 7.

Table 7. Detection limits in µg/L blood

Analyte	Detection limit [µg/L blood]
Diazinon	2.3
Fenitrothion	2.6
Malathion	2.2
Fenthion	0.5
Chlorpyrifos	2.5

9.4 Sources of error

9.4.1 Special aspects to be observed during sample processing and analysis

It is important the ensure that the Extrelut NT20 cartridges are completely dry after conditioning, as otherwise losses may occur during extraction of the organophosphates from whole blood.

Evaporation to dryness in the stream of nitrogen must be avoided at all costs. The formation of 2 phases during evaporation can be a critical step. In this case the organic phase must be withdrawn using a pipette before further processing (see Section 3.1).

The ion trace m/z 125 as the qualifier for the organophosphates fenitrothion, malathion and fenthion may suffer sensitive interference from matrix components that are not completely eliminated.

After evaporation of the toluene extract, the bromophos-methyl can be added as a chromatography standard to monitor the response and check the sensitivity and stability of the GC/MS system. If this chromatography standard shows a weak response, the GC/MS system must be checked, as cleaning of the ion source may be necessary under certain circumstances.

9.4.2 Investigations on the stability of the organophosphates in whole blood

Organophosphates were added to fresh K_2-EDTA whole blood (pooled human blood) in a concentration of 8 µg/L blood. Then the blood samples were stored for 1 h, 20 h, 44 h and 1 week respectively at 4 to 6 °C in the refrigerator, processed as described in Section 3.1 and analysed by gas chromatography/mass spectrometry according to the instructions in Section 4 and 5.
Similarly, the esterase inhibitor pyridostigmine bromide was added to fresh K_2-EDTA whole blood (111 µg/mL blood) and the blood was spiked with the organophosphates in a concentration of 8 µg/L blood in order to investigate the influence on the stability of the organophosphates in whole blood.
The cholinesterase activity (acetylcholinesterase activity of the erythrocytes, substrate: acetylthiocholine, pseudocholinesterase activity in plasma, substrate: butyrylthiocholine) was determined at each measurement time. In addition, the activities were determined in control blood (whole blood without organophosphates and without the inhibitor).
The activity of the acetylcholinesterases in control blood and in the spiked samples was between 3880 and 4590 U/L in the investigation period of 1 week, the activities of the pseudocholinesterases were in the range of 5210 to 5720 U/L, and thus in the range of normal values. After addition of the inhibitor pyridostigmine bromide, the esterase activity showed a distinct reduction to values of 610 to 1070 U/L and 110 to 130 U/L respectively.
The results for the recovery of the organophosphates to be investigated are shown in Table 8 (whole blood without addition of the inhibitor) and Table 9 (whole blood with the inhibitor pyridostigmine bromide, 111 µg/mL).
The results show that diazinon, fenitrothion and chlorpyrifos can be recovered in almost unchanged quantities, even after storage for one week in the refrigerator. Fenthion should be analysed within 47 hours. In contrast, malathion begins to be degraded during the first two days. The degradation of malathion is not inhibited by the addition of the esterase inhibitor pyridostigmine bromide. Based on these data it is not advisable to use pyridostigmine bromide to inhibit the cholinesterases before sample preparation. Alternative inhibitors were not investigated.

Table 8. Results of the recovery investigations of samples stored for up to one week

Sample	Time	Without inhibitor Recovery in % with respect to the starting sample				
		Diazinon	Fenitrothion	Malathion	Fenthion	Chlorpyrifos
0	1 h	100	100	100	100	100
1	20 h	96	101	85	94	89
2	44 h	95	116	43	91	93
3	47 h	157	92	46	92	108
4	1 week	99	108	23	75	101

Table 9. Results of the recovery investigations of samples with an added inhibitor and stored for up to one week

Sample	Time	Addition of pyridostigmine bromide (111 µg/mL) Recovery in % with respect to the starting sample				
		Diazinon	Fenitrothion	Malathion	Fenthion	Chlorpyrifos
0	1 h	100	100	100	100	100
1	20 h	107	107	83	117	108
2	44 h	123	96	53	123	115
3	47 h	113	120	49	91	68
4	1 week	116	100	0	106	117

10 Discussion of the method

The organophosphates investigated in the method described here are active substances that are present in disinfestation agents that are listed in the "*Bekanntmachung der geprüften und anerkannten Mittel und Verfahren zur Bekämpfung von tierischen Schädlingen nach § 10c Bundesseuchengesetz*" [Notification of the tested and recognised agents and procedures for the control of animal pests according to Article 10c of the German Epidemic Law] [2]. These agents are used by professional pest controllers.

A method described in the literature for the determination of toxic pesticides in whole blood [11], in which a comparable method of sample preparation and analysis are used, covers a different range of active substances. Furthermore, this method is less sensitive by a factor of about 10 than the procedure described here and is therefore more suitable for use in cases of acute intoxication.

The other methods of analysing chlorpyrifos or diazinon and chlorpyrifos in human blood and plasma samples included in the references describe the use of liquid/liquid extraction and subsequent measurement by means of GC/MS with negative chemical ionisation [12] or solid phase extraction and measurement by means of high-resolution mass spectrometry [13]. These methods are suitable for assay in the range of relevance to occupational and environmental medicine.

Using the method presented here it is possible to determine the organophosphates in human blood with good specificity, precision (rel. standard deviation <13%) and accuracy (mean recovery from 91 to 129%). The examiners of the method were able to replicate the precision, recovery and detection limits stated by the authors without problems. According to the examiners Varian Chemelut CE-1005 columns can be used as a possible alternative to the solid phase cartridges stated by the authors.

The analysis of the organophosphates in whole blood requires a relatively time-consuming clean-up, which restricts the number of samples that can be processed to only about 10 samples a day. This method seems poorly suited for routine analysis. The reasons for this include:
– The difficulty in handling the Extrelut NT-20 cartridges. However, in order to be able to use the given sample volume and to ensure sensitivity the size of the cartridge may not be reduced.

- The long chromatographic run times that are required to separate the analytes effectively from interfering matrix components.
- The samples remain relatively heavily contaminated by the matrix despite the laborious clean-up and this may lead to contamination of the ion source with more frequent cleaning cycles as a consequence.

The sensitivity of the method can be increased by the use of more powerful detectors (Agilent MSD 5973/5975) and by the injection of a larger sample volume (2 µL).

Instruments used:
Gas chromatograph HP-5890 II with mass selective detector HP-5970, autosampler 7673 and data system from Hewlett-Packard.

11 References

[1] *W. Perkow:* Wirksubstanzen der Pflanzenschutz- und Schädlingsbekämpfungsmittel. 3rd edition, Parey Buchverlag, Berlin (1999).
[2] *BgVV:* Bekanntmachungen des Bundesinstituts für gesundheitlichen Verbraucherschutz und Veterinärmedizin: Bekanntmachung der geprüften und anerkannten Mittel und Verfahren zur Bekämpfung von tierischen Schädlingen nach §10c Bundesseuchengesetz, 17th issue. Bundesgesundheitsbl. – Gesundheitsforsch. – Gesundheitsschutz 43/Suppl.2, pp. 61–74 (2000).
[3] *IPCS:* Inventory of IPCS and other WHO pesticides evaluations and summary of toxicological evaluations performed by the Joint Meeting on Pesticide Residues (1999). Available on the Internet under: *http://www.who.int/ipcs/publications/en/inventory2.pdf*
[4] *Deutsche Forschungsgemeinschaft:* List of MAK and BAT Values 2006, 42nd report, Wiley-VCH, Weinheim (2006).
[5] *H. Greim (Ed.):* Diazinon. Occupational Toxicants – Critical Data Evaluation for MAK Values and Classification of Carcinogens. Vol. 11. Wiley-VCH, Weinheim (1998).
[6] *H. Greim (Ed.):* Fenthion. Gesundheitsschädliche Arbeitsstoffe. Toxikologisch-arbeitsmedizinische Begründungen von MAK-Werten. Issues 8, 34. Wiley-VCH, Weinheim (1981, 2002).
[7] *H. Greim (Ed.):* Malathion. Gesundheitsschädliche Arbeitsstoffe. Toxikologisch-arbeitsmedizinische Begründungen von MAK-Werten. Issues 2, 18, 34, 42. Wiley-VCH, Weinheim (1973, 1992, 2002, 2007).
[8] *Bundesärztekammer:* Richtlinie der Bundesärztekammer zur Qualitätssicherung quantitativer laboratoriumsmedizinischer Untersuchungen. Dt. Ärztebl. 100, A3335 – A3338 (2003).
[9] *J. Angerer* and *G. Lehnert:* Anforderungen an arbeitsmedizinisch-toxikologische Analysen – Stand der Technik. Dt. Ärztebl. 37, C1753–C1760 (1997).
[10] *J. Angerer, Th. Göen* and *G. Lehnert:* Mindestanforderungen an die Qualität von umweltmedizinisch-toxikologischen Analysen. Umweltmed. Forsch. Prax. 3, 307–312 (1998).
[11] *T. Frenzel, H. Sochor, K. Speer* and *M. Uihlein:* Rapid multimethod for verification and determination of toxic pesticides in whole blood by means of capillary GC-MS. J. Anal. Toxicol. 24, 365–371 (2000).
[12] *K.A. Brzak, D.W. Harms, M.J. Bartels* and *R.J. Nolan:* Determination of chlorpyrifos, chlorpyrifos-oxon and 3,5,6-trichloro-2-pyridinol in rat and human blood. J. Anal. Toxicol. 22, 203–210 (1998).
[13] *D.B. Barr, J.R. Barr, V.L. Maggio, R.D. Whitehead Jr., M.A. Sadowski, R.M. Whyatt* and *L.L. Needham:* A multi-analyte method for the quantification of contemporary pesticides in human serum and plasma using high-resolution mass spectrometry. J. Chromatogr. B 778 (1–2), 99–111 (2002).

Authors: *E. Berger-Preiss, S. Gerling*
Examiners: *J. Lewalter, K. Wilmersdorf*

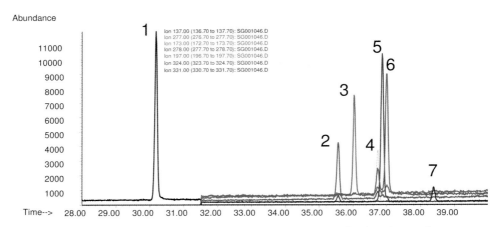

Fig. 2. Total ion chromatogram of an organophosphate standard solution in toluene after measurement by means of GC-MS (EI) in the SIM mode (1 – diazinon, 2 – fenitrothion, 3 – malathion, 4 – chlorpyrifos-diethyl-D_{10}, 5 – fenthion, 6 – chlorpyrifos, 7 – bromophos-methyl)

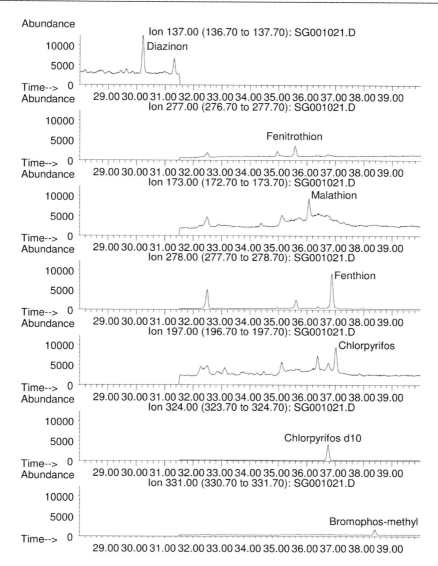

Fig. 3. GC/MS SIM chromatogram of a processed calibration standard in pooled human blood (24 µg organophosphate/L blood in each case).

Organophosphates

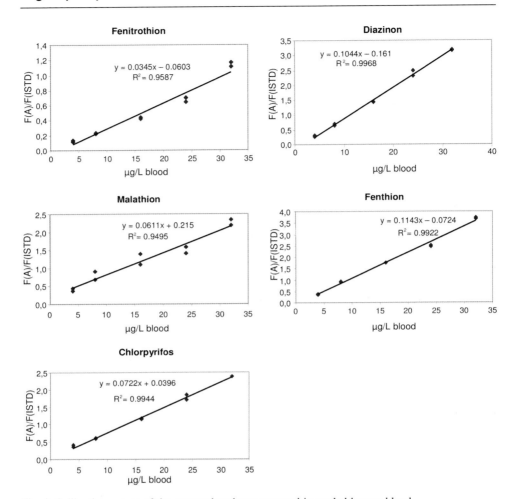

Fig. 4. Calibration curves of the organophosphates prepared in pooled human blood.

Palladium

Application Determination in urine

Analytical principle High performance liquid chromatography/ UV detection

Summary

The method described here is suitable for the quantification of palladium in urine in the concentration range of relevance to occupational medicine. For this purpose the urine is mineralised by means of UV digestion in order to minimise interference due to the organic matrix. Then the palladium is selectively converted to a complex on a solid phase (C_{18} material) using N,N-diethyl-N'-benzoylthiourea as a ligand. After elution of the complex with ethanol, it is separated by high performance liquid chromatography and registered by a UV detector. The concentration of palladium is calculated from calibration curves. The calibration curves are plotted by analysis of calibration standard solutions that are prepared in pooled urine and are processed and analysed in the same manner as the samples to be assayed.

Palladium

Within-series imprecision: Standard deviation (rel.) $s_w = 11\%$ or 5%
Prognostic range $u = 27\%$ or 12%
at a spiked concentration of 75 ng or 180 ng palladium per litre urine and where n = 6 determinations

Between-day imprecision: Standard deviation (rel.) $s_w = 11\%$ or 7%
Prognostic range $u = 25\%$ or 16%
at a spiked concentration of 75 ng or 180 ng palladium per litre urine and where n = 10 determinations

Accuracy: Recovery rate $r = 96\%$ at 75 ng/L and 94% at 180 ng/L

Detection limit: 10 ng palladium per litre urine

Palladium

The precious metal palladium is widely distributed in industrialised countries and can enter environmental compartments and the human organism by different routes. Automobiles have been fitted with catalytic converters since the beginning of the 1980s in Germany. They generally consist of a basic ceramic body coated with a layer of catalytically active noble metals. In addition to platinum and rhodium, the use of palladium has increased since 1990. As a result of the high operational temperatures and mechanical abrasion, palladium is emitted from the exhaust into the environment in its metallic form or as particulate or vaporous compounds [1].

Table 1. Overview of studies on the internal exposure to palladium

References	Method	Study size, description [n=]	Concentration Mean value; range [ng/L]
General population			
[7]	ICP-MS after dilution	14	100; 30–200
[8–10]	Sector field ICP-MS after UV digestion	21	140; 33–220
		17	52; 10–134
		17	31; 13–48
		262	31; 6–91
[11]	AAS after UV digestion and liquid/liquid extraction	9	38.7; <20–80
[12]	Sector field ICP-MS after microwave digestion	30	10; <5–21
[13]	Online FI-GFAAS after SPE	6	<36
[14]	TXRF after reductive precipitation with mercury	5	<2.5
Our own measurements	HPLC-UV after UV digestion and SPE	44	<DL; <10–28
Occupationally exposed persons			
[10]	Sector field ICP-MS after UV digestion	27 Dental technicians	135; 8–1236
[13]	Online FI-GFAAS after SPE	12 Catalytic converter industry	<80–3400
[14]	TXRF after reductive precipitation with mercury	7 Catalytic converter industry	200–1000
Our own measurements	HPLC-UV after UV digestion and SPE	6 Catalytic converter industry	405–2538
		4 Precious metal processing	<10–555

ICP=inductively coupled plasma, MS=mass spectrometry, AAS=atomic absorption spectrometry, FI-GFAAS=flow injection graphite furnace atomic absorption spectrometry, SPE=solid phase extraction, TXRF=total x-ray fluorescence analysis, HPLC=high performance liquid chromatography

A further important area of use is in inlays and crowns in dental prostheses. Dental alloys based on gold also contain up to 38.5% palladium, whereby those based on palladium contain up to almost 80% palladium [2]. The metal can enter the organism by means of mechanical abrasion and galvanic effects in the oral cavity.

Intake with food or medication is also under discussion as a third source of entry of the noble metal, as palladium residues from catalytic processes may remain after production.

Palladium also poses a problem for occupational medicine, as concentrations at the workplace, e.g. in the manufacture of catalytic converters, can be several times higher than the levels in the environment (see Table 1).

In rats, palladium taken up by the oral route is almost completely excreted in the faeces, only a very small proportion is absorbed. Similarly, the absorption rate is not very high following inhalation. The highest concentrations are found in the lungs followed by the kidneys, the spleen, the bones, and the liver. As a consequence of intravenous administration palladium is found in almost all the organs after 24 h. The major part is eliminated by the kidneys (approx. 76% after 7 days), only a small proportion is excreted in the faeces. Furthermore, palladium has been shown to be capable of passing the placental barrier [3, 4].

Experience with human exposure and information from animal studies is insufficient to enable assignment of a MAK value for palladium [5, 6]. Palladium(II) chloride and bioavailable palladium(II) compounds have been assigned to the skin-sensitising substances (Sh) by the MAK Commission. A comprehensive review of the toxicological aspects of palladium can be found in the justification for the MAK classification [6].

Table 1 shows an overview of studies on the internal exposure to palladium of the general population and of persons who are exposed to the metal at the workplace.

Authors: *J. Angerer, G. Philippeit*
Examiners: *P. Schramel, B. Michalke*

Palladium

Application   Determination in urine

Analytical principle High performance liquid chromatography/
                         UV detection

Contents

1 General principles
2 Equipment, chemicals and solutions
2.1 Equipment
2.2 Chemicals
2.3 Solutions
2.4 Calibration standards
2.5 Filling and conditioning the SPE columns
3 Specimen collection and sample preparation
3.1 Sample preparation
4 Operational parameters
4.1 Operational parameters for high performance liquid chromatography
5 Analytical determination
6 Calibration
7 Calculation of the analytical result
8 Standardisation and quality control
9 Evaluation of the method
9.1 Precision
9.2 Accuracy
9.3 Detection limits
9.4 Sources of error
10 Discussion of the method
11 References

1 General principles

The urine is mineralised by means of UV digestion in order to minimise interference due to the organic matrix. Then the palladium is selectively converted to a complex on a solid phase (C_{18} material) using N,N-diethyl-N'-benzoylthiourea (DEBT) as a ligand. After elution of the complex with ethanol, it is separated by high performance liquid chromatography and registered by a UV detector. The concentration of palladium is calculated from calibration curves. The calibration curves are plotted by analysis of calibration standard solutions that are prepared in pooled urine and are processed and analysed in the same manner as the samples to be analysed.

2 Equipment, chemicals and solutions

2.1 Equipment

HPLC-UV system consisting of an isocratic pump, an injection valve, a UV detector suitable for measurement at a wavelength of 274 nm, and an integrator or PC system for data processing.

HPLC column:
Lichrospher® RP 18e, length: 250 mm; inner diameter: 4 mm; particle diameter: 5 µm (e.g. VWR, No. 1.50838.0001)

Precolumn:
Lichrospher® RP 18e, length: 4 mm; inner diameter: 4 mm; particle diameter: 5 µm (e.g. VWR, No. 1.50962.0001)

250 mL Plastic bottles

250 mL Glass beaker

UV digester (e.g. 705, Metrohm)

14 mL Quartz glass vessels and PTFE stoppers for the UV digester (e.g. from Metrohm)

Thermosensor (e.g. Omron E5CS, Metrohm)

Solid phase extraction station (e.g. VacElut, Varian)

Syringe for injection into the HPLC (e.g. Gastight® No. 1710, Hamilton)

10, 100 and 1000 mL Volumetric flasks

100 mL and 1000 mL Measuring cylinders

1000 mL Glass bottle

15 mL Plastic centrifuge tubes (e.g. from Sarstedt)

Empty 4 mL polyethylene reservoirs for solid phase enrichment (e.g. from ICT Handels-GmbH)

Polyethylene frits (20 µm, diameter 9 mm) (e.g. from ICT Handels-GmbH)

1.5 mL Reaction vessel (e.g. from Eppendorf)

Adjustable pipettes (1–100 µL, 100–1000 µL) (e.g. Varipettes®, Eppendorf)

Multipipettes (e.g. Multipette®, Eppendorf)

Ultrasonic bath (e.g. Bransonic 220, Sigma-Aldrich)

Water jet pump

2.2 Chemicals

Anhydrous acetic acid, p.a. (100%, glacial acetic acid) (e.g. Merck, No. 1.00063.1000)

Sulphuric acid, p.a. (95–97%) (e.g. Merck, No. 1.00731.1000)

Hydrogen peroxide Suprapur® (30%) (e.g. Merck, No. 1.07298.1000)

Nitric acid Suprapur® (65%) (e.g. Merck, No. 1.07298.1000)

N,N-diethyl-N'-benzoylthiourea (prepared according to reference [15])

Ethanol LiChrosolv® (e.g. Merck, No. 1.11727.1000)

Methanol LiChrosolv® (e.g. Merck, No. 1.06018.2500)

Palladium standard solution 1.0 g/L in 5% nitric acid (e.g. Sigma-Aldrich, No. 76035-100ML)

Octadecyl(C_{18}) material (40–60 μm, 120 Å) for SPE (e.g. Bakerbond®, Mallinckrodt Baker, No. 7488-05)

Ultrapure water (e.g. Milli-Q water, Millipore)

2.3 Solutions

1.2 M Nitric acid:
About 500 mL ultrapure water are placed in a 1000 mL volumetric flask. 82 mL of 65% nitric acid are measured in a measuring cylinder and added to the volumetric flask. The flask is subsequently filled to its nominal volume with ultrapure water. The 1.2 M nitric acid is stable for at least 6 months when tightly sealed and stored at room temperature.

N,N-diethyl-N'-benzoylthiourea (DEBT solution, 50 mg/L):
5 mg N,N-diethyl-N'-benzoylthiourea are weighed in a 100 mL volumetric flask and dissolved in 1 mL methanol. The flask is subsequently filled to its nominal volume with ultrapure water. This solution must be freshly prepared every day.

DEBT solution (5 mg/L):
Approx. 50 mL of 1.2 M nitric acid are placed in a 100 mL volumetric flask and 10 mL of the 50 mg/L DEBT solution are added using a pipette. The flask is subsequently filled to its nominal volume with 1.2 M nitric acid. This solution must be freshly prepared every day.

Mobile phase (methanol/water, 98:2 (v/v)):
980 mL Suprasolv® methanol are measured using a measuring cylinder and then transferred to a 1000 mL glass bottle. Then 20 mL ultrapure water are added using a pipette. The mobile phase is thoroughly mixed and degassed for a time in the ultrasonic bath. This solution is stable at room temperature for at least 4 weeks.

2.4 Calibration standards

Starting solution for the calibration standard:
A commercially available palladium standard solution in nitric acid (1 g/L) serves as the starting solution.

Stock solution:
Approx. 50 mL of the 1.2 M nitric acid are placed in a 100 mL volumetric flask and 1 mL of the palladium starting solution is added using a pipette. The flask is subsequently filled to its nominal volume with 1.2 M nitric acid (10 mg/L).
This solution is stable for at least 3 months when stored at +4 to +6 °C.

Working solution:
Approx. 50 mL of the 1.2 M nitric acid are placed in a 100 mL volumetric flask and 100 µL of the palladium stock solution (10 mg/L) are added using a pipette. The flask is subsequently filled to its nominal volume with 1.2 M nitric acid (10 µg/L).
This solution must be freshly prepared every day.

The calibration standard solutions are prepared in a 10 mL volumetric flask by dilution of the palladium working solution with pooled urine from persons who have not been exposed to the metal. The pipetting scheme for the preparation is shown in Table 2. The solutions are subsequently shaken vigorously and then processed and analysed as described in the following sections. These calibration standard solutions must be freshly prepared every day.

Table 2. Pipetting scheme for the preparation of the calibration standard solutions

Volume of the working solution [µL]	Volume of the pooled urine [mL]	Concentration of the standard solution [ng/L]
0	10	0
50	9.95	50
100	9.9	100
150	9.85	150
200	9.8	200
250	9.75	250
500	9.5	500

2.5 Filling and conditioning the SPE columns

A frit is placed in each of the empty reservoirs. 80 mg of the C_{18} material are weighed in. Another frit is placed on the C_{18} material. The filled columns are mounted on the solid phase extraction station. The columns are conditioned first with 3 mL methanol, then with 3 mL of 1.2 M nitric acid.

3 Specimen collection and sample preparation

The urine is collected in plastic bottles, acidified with glacial acetic acid (1 mL to 100 mL urine) and stored at $-20\,°C$ until processing. The urine samples are stable for at least 1 year when stored under these conditions.

In order to determine any reagent blank values 10 mL water are included in each analytical series instead of urine, and the water is subjected to processing and analysis as described in the following sections.

3.1 Sample preparation

The urine sample must be carefully homogenised by intensive shaking before sample processing is carried out. If any sediment has formed, the sample is warmed in a glass beaker to dissolve the sediment. If necessary, the sample must be vigorously shaken before being divided into aliquots to ensure that any residual precipitate is distributed as homogeneously as possible. 10 mL urine, 500 µL sulphuric acid and 500 µL hydrogen peroxide are pipetted into a 14 mL quartz glass vessel, which is then sealed with a PTFE stopper. The glass vessels are placed in the UV digester and irradiated with UV light for 6 hours at approx. $90\,°C$. The temperature of the digester is regulated externally by means of a thermosensor and a circulation loop of cooling water. During the irradiation period 500 µL hydrogen peroxide are added 4 to 5 times at regular intervals. If the samples still show a yellowish coloration after 6 hours, hydrogen peroxide is added again and irradiation is continued until the solution is clear. The digested sample is brought to room temperature in the quartz glass vessel overnight and transferred to a plastic centrifuge tube the next day.

After being filled and conditioned as described in Section 2.5, the SPE column is loaded with 3 mL of the DEBT solution (5 mg/L). When the solution has seeped through, the entire digestion solution is gradually applied to the column. The palladium is selectively enriched by formation of a complex with the DEBT ligand that has been adsorbed by the C_{18} material. A neutral complex $Pd(DEBT)_2$ is formed in this reaction (see Figure 1). After the digestion solution has drained off completely, the column material is dried for approx. 5 min by applying a vacuum. The $Pd(DEBT)_2$ complex is then eluted from the column into a 1.5 mL reaction vessel using 450 µL ethanol. If no eluate emerges, a slight vacuum can be briefly applied to the SPE station. The eluate should be analysed on the same day if possible.

4 Operational parameters

4.1 Operational parameters for high performance liquid chromatography

The analytical measurement is carried out on an HPLC-UV system consisting of an isocratic pump, an injection valve, a UV detector capable of measurement at a wavelength of 274 nm, and an integrator or PC system for data processing.

Separation column:	Material:	Steel
	Length:	250 mm
	Inner diameter:	4 mm
	Column packing:	LiChrospher® RP 18e (5 µm)
Precolumn:	Material:	Steel
	Length:	4 mm
	Inner diameter:	4 mm
	Column packing:	LiChrospher® RP 18e (5 µm)
Separation principle:	Reversed phase	
Temperature:	Room temperature	
Detection:	UV detector, 274 nm, attenuation 0.5	
Mobile phase:	Methanol/water = 98:2 (v/v)	
	(The mobile phase must be degassed before use)	
Gradient:	Isocratic	
Stop time:	10 minutes	
Flow rate:	1.0 mL/min	
Injection volume:	200 µL	

All other parameters must be optimised in accordance with the manufacturer's instructions.

5 Analytical determination

200 µL of the eluate obtained as described in Section 3.1 are injected into the HPLC by means of a syringe. A reagent blank and a quality control solution are included in each analytical series. If the measured values are not within the linear range of the calibration curve, the urine samples are appropriately diluted with ultrapure water and processed anew.

A retention time (RT) of 6.9 minutes was found for the $Pd(DEBT)_2$ complex under the chromatographic conditions described here. The signal for the palladium complex is clearly separated from the signal of the free ligand (RT between 2 and 3 min). These retention times serve only as guidelines. Users of the method must satisfy themselves of the separation power of the HPLC column they use and of the resulting retention behaviour of the analyte.

Figure 2 shows chromatograms of processed urine samples. In this case a sample from a person who was not exposed to palladium and that of a person who was subjected to occupational exposure are given as examples. Slight shifts in the retention time can be explained by the fact that injection into the HPLC and starting of the integrator were carried out manually.

6 Calibration

The calibration standard solutions prepared in urine as described in Section 2.4 are processed in the same manner as the urine samples prepared according to Section 3.1 and analysed in accordance with the instrumental parameters listed in Section 4.1. In each case 200 µL of the processed calibration standard solutions are injected into the HPLC.

The linear calibration graph is obtained by plotting the peak areas of the $Pd(DEBT)_2$ complex as a function of the palladium concentrations used. Any reagent blank values and any background exposure revealed in the urine that is used to prepare the calibration standard solutions must be taken into account and subtracted from the results. It is not necessary to plot a complete calibration graph for every analytical series. Inclusion of a single standard solution in urine should be sufficient. If major deviations from the values in the linear calibration graph are recorded for this standard, then the reasons must be clarified. If the quality control sample also exhibits significant deviations, the entire linear calibration curve may have to be plotted anew. The calibration graph is linear between the detection limit and 500 ng palladium per litre urine.

Figure 3 shows an example of a linear calibration graph for palladium in human urine.

7 Calculation of the analytical result

The peak areas obtained for palladium in a urine sample are used to read off the appropriate concentration in ng per litre urine from the relevant linear calibration graph. Any reagent blank values must be subtracted from the analytical results for the real samples.

If the urine sample has been previously diluted, this must be taken into account when the result is calculated by multiplying the analytical result by the dilution factor.

8 Standardisation and quality control

Quality control of the analytical results is carried out as stipulated in the guidelines of the Bundesärztekammer (German Medical Association) [16] and in the special preliminary remarks to this series. No commercially available control material is available for internal quality assurance. Therefore control material must be prepared in the laboratory. For this purpose pooled urine is spiked with a defined quantity of palladium. The concentration of this control material should be in the concentration range of the background exposure of the general population (see Table 1). A six-month supply of this control material is prepared, divided into aliquots in suitable vessels and stored in the deep-freezer. The control material is prepared freshly every six months. The expected value and the tolerance range of this quality control material are ascertained in a pre-analytical period (one analysis of the control material on each of 20 different days) [17–19].

9 Evaluation of the method

9.1 Precision

Pooled urine was spiked with palladium concentrations of 75 ng/L and 180 ng/L, and then processed and analysed as described in the previous sections to check the precision in the series. Six replicate analyses of these urine samples yielded the precision in the series shown in Table 3. In addition, the precision from day to day was determined. The same spiked pooled urine is used for this purpose. This urine was processed and analysed on 10 different days.

Table 3. Precision for the determination of palladium

	n	Spiked concentration [ng/L]	Standard deviation (rel.) [%]	Prognostic range [%]
In the series	6	75	11	27
		180	5	12
From day to day	10	75	11	25
		180	7	16

9.2 Accuracy

Recovery experiments were carried out to test the accuracy of the method. For this purpose pooled urine spiked with a defined quantity of palladium was analysed 10 times. The added amount of palladium was once again 75 ng/L and 180 ng/L. The relative recovery rates were 92% and 96%.

9.3 Detection limits

Based on three times the ratio of the signal to the analytical background noise in the temporal vicinity of the analytical signal, a detection limit of 10 ng of palladium per litre urine was determined under the conditions for sample processing (UV digestion and solid phase extraction) and the conditions for HPLC-UV system given here.

9.4 Sources of error

As described in Section 2.3 it is important to ensure that methanol and water are measured separately when mixing the mobile phase, otherwise the correct mixture ratio cannot be guaranteed (volume contraction).

A UV digestion procedure preceded the analytical run to minimise interference. This leads to the destruction of the dissolved organic components. It is of critical importance that digestion is complete (as described in Section 3.1). When an insufficiently digested sample is enriched, the C_{18} material of the solid phase takes on a yellowish colour. After injection of the ethanolic eluate, which is also yellowish, the background noise is strongly increased in the chromatogram. A palladium signal is no longer recognisable in such cases (Figure 4). Complete destruction of the organic components, which are also strongly adsorbed by the stationary phase, avoids overloading of the solid phase material during enrichment. Moreover, complete UV digestion is important in order to be able to minimise the matrix influence on the slope of the linear calibration graphs. Calibration was carried out in 1.2 M nitric acid as well as urine to check this influence. Almost identical slopes were achieved for calibration in the matrix and in nitric acid, and the calibration points varied within similar confidence limits in both cases (Figure 3). Despite these findings, matrix calibration is preferable, as recovery rates deteriorated in the case of calibration with nitric acid at low palladium concentrations in some cases during the examination of the method.

The HPLC-UV analysis is performed at room temperature. In the summer months it was observed that the retention times were reduced, leading to a shift in the Pd(DEBT)$_2$ signals as a result of higher room temperatures. The consequence was that the analyte signal could not be clearly separated from the relatively large signal generated by the free ligand. However, the examiners of the method did not ascertain such effects. If such effects are evident, the analytical column can be cooled using refrigeration bricks. This normalises the retention behaviour of the components again. It is advisable to thermostatically control the temperature of the HPLC column using a column oven.

10 Discussion of the method

For many years determination of palladium posed a seemingly insoluble analytical problem. The highly sensitive method of inductively coupled plasma with mass selective detection (ICP-MS) that is frequently used in inorganic trace analysis failed to determine palladium. Spectral interference overlapped the palladium signal and false positive values were recorded [7, 8]. It is only in the last three years as a result of

Table 4. Overview of the analytical methods described in the literature for the determination of palladium and the reliability criteria

References	Method	V_{sample} [mL]	DL [ng/L]	Precision in the series VC	Rel. recovery
[7]	ICP-MS after dilution	8.9	30	At 192 ng/L: 1.5% (n=10)	At 2000 ng/L: 87.5%
[8]	Sector field ICP-MS after UV digestion	20	0.24	No data	No data
[11]	AAS after UV digestion and liquid/liquid extraction	10	20	At 5 ng/L: 5.2% (n=10)	No data
[12]	Sector field ICP-MS after UV photolysis, or after microwave digestion, and dilution	5	0.25 or 0.1	At 5 ng/L: 2.4% (n=10)	At 20–50 ng/L: 95–98%
[13]	Online FI-GFAAS after SPE	2.7	36	At 100 ng/L aq: 4.2% (n=5) At 50 ng/L aq: 5% (n=5)	No data
[14]	TXRF after high-pressure digestion and reductive precipitation with mercury	20	2.5	At 203 ng/L: 4% (n=5)	At 5–10 ng/L: >95%
[20]	Isotope dilution ICP-MS after chromatographic separation	0.2 g	0.075 ng/g	No data	No data
This method	HPLC-UV after UV digestion and SPE	10	10	At 75 ng/L: 11% (n=6) At 180 ng/L: 5% (n=6)	At 75 ng/L: 94% At 180 ng/L: 96%

V_{sample} = volume of sample used, DL = detection limit, VC = variation coefficient, ICP = inductively coupled plasma, MS = mass spectrometry, AAS = atomic absorption spectrometry, FI-GFAAS = flow injection graphite furnace atomic absorption spectrometry, SPE = solid phase extraction, TXRF = total x-ray fluorescence analysis, HPLC = high performance liquid chromatography.

great endeavours in the field of analysis that detection methods suitable for analysing the background exposure of the general population to palladium have been developed [12–14, 20]. Table 4 summarises the analytical methods for the determination of palladium described in the literature.

The method described here is a sensitive and powerful analytical method for the determination of palladium in human urine. The sensitivity is achieved by means of selective formation of a palladium complex with the ligand *N,N*-diethyl-*N'*-benzoylthiourea (DEBT), which is specific for the metal, and the effective enrichment of the resulting metal complex on the C_{18} solid phase material. According to Douglass and Dains, DEBT is synthesised from benzoylisothiocyanate and 2-aminoethanol, followed by recrystallisation with ethanol [15]. Until recently DEBT was commercially available, e.g. from Fluka. The use of DEBT as a complexing agent for palladium is based on the work of Schuster and Schwarzer [13]. The method described here is notable for the fact that complex formation takes place on the solid phase material. To achieve this the modified C_{18} material is loaded with an excess of DEBT. Then the sample material is applied to the column. Palladium reacts rapidly with DEBT at room temperature. This property is exploited to separate palladium from platinum, which is also of interest to environmental medicine. Although DEBT is also a selective ligand for platinum, higher temperatures are necessary (approx. 60 °C) for formation of the complex. As both the free DEBT and the palladium complex are strongly adsorbed by the non-polar C_{18} material, the losses of palladium due to processing are extremely slight. The Pd(DEBT)$_2$ complex is readily soluble in ethanol. A volume of 450 µL is sufficient for complete elution. After thorough cleaning with methanol among other agents, the SPE reservoirs can be filled and used again as described in Section 2.5.

The separation of the Pd(DEBT)$_2$ complex from the free, excess ligand DEBT, which is also readily soluble in ethanol, is performed on an analytical HPLC column filled with end-capped C_{18} material. In contrast to conventionally modified C_{18} material, end-capped C_{18} material no longer has reactive sites. This packing was used in order to prevent any losses on the analytical column. A mixture of methanol/water in the ratio of 98:2 (v/v) was used as the mobile phase. Although the organic proportion seems relatively high, good separation of the Pd(DEBT)$_2$ complex from the free ligand was achieved. Higher proportions of water (even 5%) cause the retention time of the Pd signal to shift beyond 10 min and the signal broadens. This has a negative effect on the detection limit that can be achieved.

The fact that the metal complex resulting from the conversion of palladium with DEBT has a high molar extinction coefficient ε and is therefore suitable for sensitive quantification by means of UV detection was decisive for the development of this procedure.

Seventeen complete linear calibration graphs were plotted as part of the method validation. Figure 3 shows these linear calibration graphs, in pooled urine in one case and in 1.2 M nitric acid in the other. It is evident from Figure 3 that the gradient of the linear calibration graphs varies only within narrow limits. Therefore, as a rule, a one-point calibration as described in Section 6 is sufficient.

The reliability of this newly developed method was investigated under the conditions given for sample processing and HPLC-UV. Both the precision in the series and the precision from day to day with values below 11% and the relative recovery rates of 94 to 96% at concentrations of 75 and 180 ng palladium per litre urine can be regarded as good. Based on three times the signal/background noise ratio, a detection limit of 10 ng Pd/L was determined for the HPLC-UV method. Comparison of our reliability data with the reliability criteria given for methods described in the literature also showed our results to be very satisfactory (see Table 4).

The procedure described here is reproducible, specific and sensitive. The method can be regarded as practicable, as relatively modest instrumental resources are needed and it is suitable for routine use.

Instruments used:
LaChrom L-7110 isocratic pump (Merck-Hitachi), Vista 5500 HPLC system (Varian), L-4000 UV detector (Merck-Hitachi) and 4290 integrator (Varian)

11 References

[1] F. Zereini, F. Alt, K. Rankenburg, J. M. Beyer and S. Artelt: The distribution of platinum group elements (PGE) in the environmental compartments of soil, mud, roadside dust, road sweepings and water: Emission of platinum group elements (PGE) from motor vehicle catalytic converters. UWSF – Z. Umweltchem. Ökotox. 9, 193–200 (1997).
[2] DeguDent: Gesamtübersicht Dentallegierungen. In: http://www.degudent.de/ Kommunikation_und_Service/Download/Legierungen/Direktvertrieb/Wandtabelle_deutsch.pdf.
[3] W. Moore, D. Hysell, W. Crocker and J. Stara: Biological Fate of ^{103}Pd in rats following different routes of exposure. Environ. Res. 8(2), 234–240 (1974).
[4] W. Moore, D. Hysell, L. Hall, K. Campbell and J. Stara: Preliminary studies on the toxicity and metabolism of palladium and platinum. Environ. Health Perspect. 10, 63–71 (1975).
[5] Deutsche Forschungsgemeinschaft: MAK- und BAT-Werte-Liste 2006, Issue 42, Wiley-VCH, Weinheim (2006).
[6] H. Greim (ed.): Palladiummetall. Gesundheitsschädliche Arbeitsstoffe. Toxikologisch-arbeitsmedizinische Begründungen von MAK-Werten. Issues 33, 38, Wiley-VCH, Weinheim (2001, 2004).
[7] P. Schramel, I. Wendler and J. Angerer: The determination of metals (antimony, bismuth, lead, cadmium, mercury, palladium, platinum, tellurium, thallium, tin and tungsten) in urine samples by inductively coupled plasma-mass spectrometry. Int. Arch. Occup. Environ. Health 69(3), 219–223 (1997).
[8] J. Begerow, M. Turfeld and L. Dunemann: Determination of physiological palladium and platinum levels in urine using double focusing magnetic sector field ICP-MS. Fresenius J. Anal. Chem. 359, 427–429 (1997).
[9] J. Begerow, G. A. Wiesmüller, M. Turfeld and L. Dunemann: Welchen Beitrag liefern Emissionen aus dem Straßenverkehr zur Hintergrundbelastung der Bevölkerung mit Platin und Palladium? Poster at the 2nd meeting of the International Society of Environmental Medicine in Gießen (1998).
[10] J. Begerow, U. Sensen, G. A. Wiesmüller and L. Dunemann: Internal platinum, palladium and gold exposure of environmentally and occupationally exposed persons. Zbl. Hyg. Umweltmed. 202, 411–424 (1998/1999).
[11] J. Begerow, M. Turfeld and L. Dunemann: Determination of physiological noble metals in human urine using liquid-liquid extraction and Zeeman electrothermal atomic absorption spectrometry. Anal. Chim. Acta 340, 277–283 (1997).
[12] M. Krachler, A. Alimonti, F. Petrucci, K. J. Irgolic, F. Forestiere and S. Caroli: Analytical problems in the determination of platinum-group metals in urine by quadrupole and magnetic sector field inductively coupled plasma mass spectrometry. Anal. Chim. Acta 363, 1–10 (1998).
[13] M. Schuster and M. Schwarzer: Selective determination of palladium by on-line column preconcentration and graphite furnace atomic absorption spectrometry. Anal. Chim. Acta 328, 1–11 (1996).
[14] J. Messerschmidt, A. von Bohlen, F. Alt and R. Klockenkamper: Separation and enrichment of palladium and gold in biological and environmental samples, adapted to the determination by total reflection X-ray fluorescence. Analyst 125(3), 397–399 (2000).
[15] I. B. Douglass and F. B. Dains: Some Derivatives of Benzoyl and Furoyl Isothiocyanates and their Use in Synthesizing Heterocyclic Compounds. J. Am. Chem. Soc. 56, 719–721 (1934).
[16] Bundesärztekammer. Richtlinie der Bundesärztekammer zur Qualitätssicherung quantitativer laboratoriumsmedizinischer Untersuchungen. Dt. Ärztebl. 100, A3335–A3338 (2003).
[17] J. Angerer and G. Lehnert. Anforderungen an arbeitsmedizinisch-toxikologische Analysen – Stand der Technik. Dt. Ärztebl. 37, C1753–C1760 (1997).

[18] *J. Angerer, Th. Göen* and *G. Lehnert:* Mindestanforderungen an die Qualität von umweltmedizinisch-toxikologischen Analysen. Umweltmed. Forsch. Prax. 3, 307–312 (1998).
[19] *G. Lehnert, J. Angerer* and *K. H. Schaller:* Statusbericht über die externe Qualitätssicherung arbeits- und umweltmedizinisch-toxikologischer Analysen in biologischen Materialien. Arbeitsmed. Sozialmed. Umweltmed. 33(1), 21–26 (1998).
[20] *M. Müller* and *K. G. Heumann:* Isotope dilution inductively coupled plasma quadrupole mass spectrometry in connection with a chromatographic separation for ultra trace determinations of platinum group elements (Pt, Pd, Ru, Ir) in environmental samples. Fresenius J. Anal. Chem. 368(1), 109–115 (2000).

Authors: *J. Angerer, G. Philippeit*
Examiners: *P. Schramel, B. Michalke*

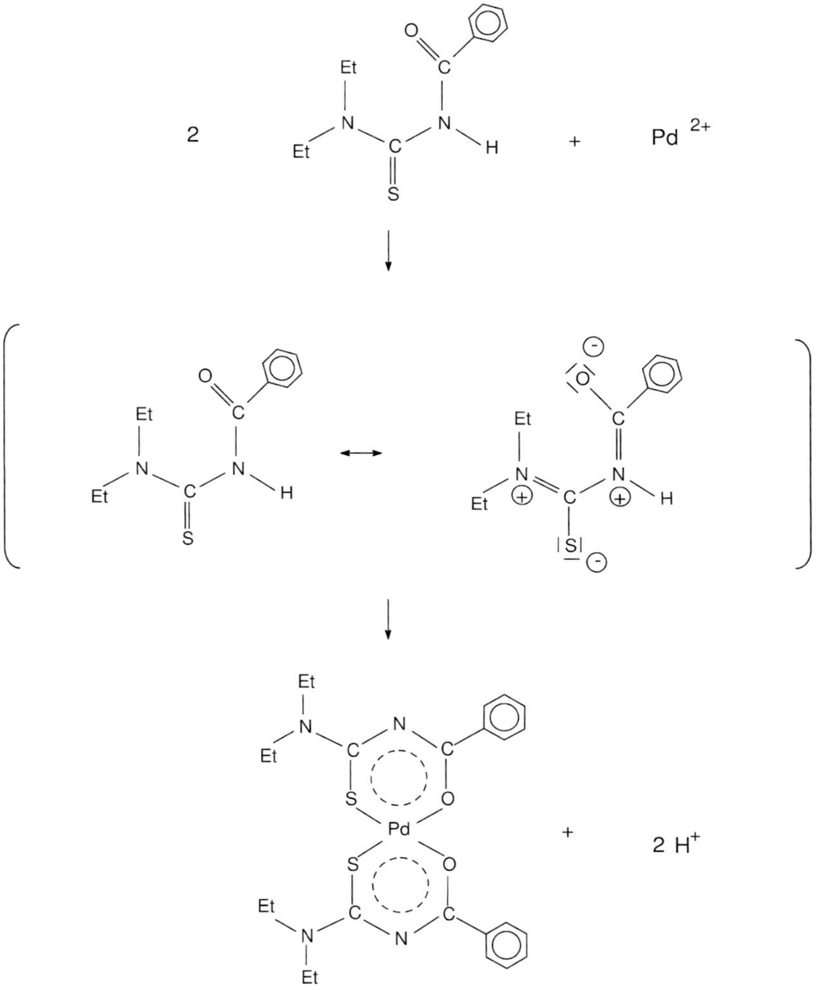

Fig. 1. Formation of a neutral complex between *N,N*-diethyl-*N'*-benzoylthiourea and palladium(II)

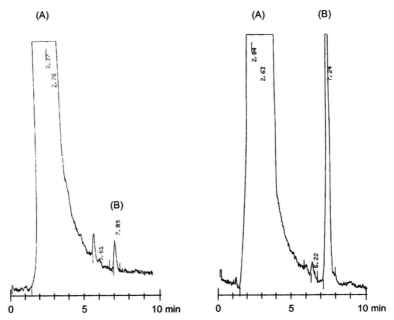

Fig. 2. Left: chromatogram of a processed urine sample from a person who was not exposed to palladium (28 ng/L). Right: chromatogram of a processed urine sample from a person who was occupationally exposed to palladium (2538 ng/L). (A) is the free ligand DEBT and (B) the analyte peak of Pd(DEBT)$_2$ in each case.

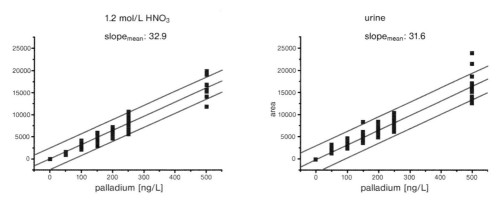

Fig. 3. Comparison of the linear calibration graphs for palladium in 1.2 M nitric acid (left) and in urine (right), with n = 17 in each case

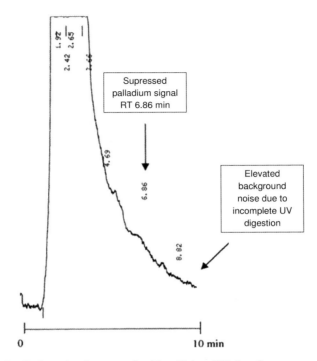

Fig. 4. Suppressed palladium signal as a result of insufficient UV digestion

Polycyclic musk compounds

Application Determination in blood

Analytical principle Capillary gas chromatography/
tandem mass spectrometric detection (GC/MS/MS)

Summary

The method described here permits the assay of the polycyclic musk compounds (PMC)   1,3,4,6,7,8-hexahydro-4,6,6,7,8,8-hexamethylcyclopenta[g]-2-benzopyrane (HHCB), 7-acetyl-1,1,3,4,4,6-hexamethyltetrahydronaphthalene (AHTN), 4-acetyl-1,1-dimethyl-6-*tert*-butyldihydroindene (ADBI) and 6-acetyl-1,1,2,3,3,5-hexamethyl-dihydroindene (AHDI) in the blood of persons exposed to these compounds in the environment.
The PMCs are transferred to an organic solvent by means of liquid-liquid extraction. Then the analytes are enriched using a combination of three consecutive solid phases in series and separated from the matrix components. After elution of the analytes, the volume of the eluate is reduced by evaporation. The analytes are then separated by capillary gas chromatography and detected by tandem mass spectrometry. AHTN with 3-fold deuterium labelling, which is used as the internal standard, is added to the sample before processing. Calibration is carried out using calibration standard solutions prepared in hexane and analysed parallel to the processed samples.

HHCB

Within-series imprecision: Standard deviation (rel.) $s_w = 19.7\%$
                          Prognostic range $u = 44.6\%$
                          at a spiked concentration of 1 µg HHCB per litre blood
                          and where n = 9 determinations

Between-day imprecision: Standard deviation (rel.) $s_w = 20.0\%$
                          Prognostic range $u = 45.2\%$
                          at a spiked concentration of 1 µg HHCB per litre blood
                          and where n = 9 determinations

Accuracy: Recovery rate $r = 115\%$

Quantitation limit: 0.32 µg HHCB per litre blood

AHTN

Within-series imprecision: Standard deviation (rel.) $s_w = 14.8\%$
Prognostic range $u = 33.5\%$
at a spiked concentration of 1 µg AHTN per litre blood
and where n = 9 determinations

Between-day imprecision: Standard deviation (rel.) $s_w = 17.3\%$
Prognostic range $u = 39.1\%$
at a spiked concentration of 1 µg AHTN per litre blood
and where n = 9 determinations

Accuracy: Recovery rate $r = 103\%$

Quantitation limit: 0.38 µg AHTN per litre blood

ADBI

Within-series imprecision: Standard deviation (rel.) $s_w = 14.9\%$
Prognostic range $u = 33.7\%$
at a spiked concentration of 1 µg ADBI per litre blood
and where n = 9 determinations

Between-day imprecision: Standard deviation (rel.) $s_w = 12.4\%$
Prognostic range $u = 28.0\%$
at a spiked concentration of 1 µg ADBI per litre blood
and where n = 9 determinations

Accuracy: Recovery rate $r = 115\%$

Quantitation limit: 0.27 µg ADBI per litre blood

AHDI

Within-series imprecision: Standard deviation (rel.) $s_w = 8.7\%$
Prognostic range $u = 19.7\%$
at a spiked concentration of 1 µg AHDI per litre blood
and where n = 9 determinations

Between-day imprecision: Standard deviation (rel.) $s_w = 16.0\%$
Prognostic range $u = 36.2\%$
at a spiked concentration of 1 µg AHDI per litre blood
and where n = 9 determinations

Accuracy: Recovery rate $r = 97\%$

Quantitation limit: 0.25 µg AHDI per litre blood

Polycyclic musk compounds

Natural musk is very expensive and its availability is limited. Therefore synthetic musk compounds are used as fragrances for washing powders, cleansing agents, cosmetics, perfumes, air freshener sprays, insect repellent sprays, textiles, etc. These synthetic musk compounds belong to two chemically separate classes of substance, i.e. the nitro musk compounds and the polycyclic musk compounds (PMC). Whereas the production and application of the nitro musk compounds is gradually decreasing, the use of PMCs has been steadily rising since the middle of the 1980s [1]. In 1995 approx. 15.5 mg of PMCs were used per head of population per day in Europe [2]. The most important compounds are HHCB, AHTN, ADBI and AHDI (see Figure 1 and Table 1).

Table 1. Chemical names, trade names and CAS numbers of the PMCs

Substance	Chemical name	Trade name	CAS number
HHCB	1,3,4,6,7,8-Hexahydro-4,6,6,7,8,8-hexamethylcyclopenta-[g]-2-benzopyrane	Galaxolide® Abbalide® Pearlide®	1222-05-5
AHTN	7-Acetyl-1,1.3.4.4.6-hexamethyltetrahydronaphthalene	Tonalide® Fixolide®	1506-02-1
ADBI	4-Acetyl-1,1-dimethyl-6-tert-butyldihydroindene	Celestolide® Crysolide®	13171-00-1
AHDI	6-Acetyl-1,1,2,3,3,5-hexamethyldihydroindene	Phantolide®	15323-35-0

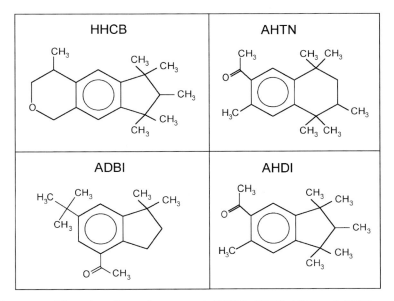

Fig. 1. Structures of the polycyclic musk compounds HHCB, AHTN, ADBI and AHDI

In 1996 approx. 5600 t of HHCB and AHTN were produced annually. These two compounds amount to approx. 70% of the global trade in synthetic musk compounds [3]. HHCB and AHTN make up more than 95% of the PMCs [2].

PMCs from cosmetics and freshly washed clothes penetrate the skin, but they also enter the human organism with food [4, 5]. They have already been detected in body fat and in mother's milk [5, 6, 7].

Angerer and Käfferlein have published a method for the analysis of nitro musk compounds in human plasma [8]. On account of the growing significance of the PMCs a method was also required to determine this substance group in blood. In 1999 Bauer and Frössl presented the first results on the occurrence of PMCs in whole blood. The mean value for HHCB was 0.722 µg/L, and for AHTN it was 0.274 µg/L (n = 328 test subjects), the corresponding 95th percentile was given as 1.651 µg/L and 0.545 µg/L respectively [9]. No significant differences were found between the concentrations for men and women. Unfortunately, Bauer and Frössl provided no information on the analysis except to state that the samples were measured "by GC-MS after solid phase extraction and evaporation to reduce the volume".

The analytical method presented here is based on the preliminary work of Eschke et al., who reported that the sensitivity of the analysis can be greatly enhanced and the "background noise" can be drastically reduced when the GC-MS/MS procedure is selected instead of the original GC-MS method [6].

Authors: *W. Butte, A. Schmidt*
Examiner: *H.-W. Hoppe*

Polycyclic musk compounds

Application   Determination in blood

Analytical principle Capillary gas chromatography/
                         tandem mass spectrometric detection (GC/MS/MS)

Contents

1 General principles
2 Equipment, chemicals and solutions
2.1 Equipment
2.2 Chemicals
2.3 Solutions
2.4 Calibration standards
2.5 Conditioning of the SPE columns
3 Specimen collection and sample preparation
3.1 Sample preparation
4 Operational parameters
4.1 Operational parameters for gas chromatography and mass spectrometry
5 Analytical determination
6 Calibration
7 Calculation of the analytical result
8 Standardisation and quality control
9 Evaluation of the method
9.1 Precision
9.2 Accuracy
9.3 Detection limits
9.4 Sources of error
10 Discussion of the method
11 References

1 General principles

The polycyclic musk compounds are transferred to an organic solvent by means of liquid-liquid extraction. Then the analytes are processed using a combination of SPE cartridges consisting of silica gel-sulphuric acid/benzenesulphonic acid resin and silica gel. This destroys the fat in blood and removes the matrix components. After elution, the analytes are separated by capillary gas chromatography and detected by tan-

The MAK-Collection Part IV: Biomonitoring Methods, Vol. 11.
DFG, Deutsche Forschungsgemeinschaft
Copyright © 2008 WILEY-VCH Verlag GmbH & Co. KGaA, Weinheim
ISBN: 978-3-527-31596-3

dem mass spectrometry. AHTN with 3-fold deuterium labelling, which is used as the internal standard, is added to the sample before processing. Calibration is carried out with calibration standard solutions prepared in hexane and analysed parallel to the processed samples.

2 Equipment, chemicals and solutions

2.1 Equipment

Gas chromatograph with split/splitless injector, autosampler, ion-trap mass spectrometer with the tandem MS option and data processing system

Capillary gas chromatographic column:
Length: 30 m; inner diameter: 0.25 mm; stationary phase: cross-linked methyl-phenyl-polysiloxane block polymers; film thickness: 0.25 µm (e.g. Macherey-Nagel OPTIMA® δ-3, No. 726420.30)

Disposable syringes containing an anticoagulant (e.g. potassium EDTA Monovettes from Sarstedt)

Centrifuge (e.g. from Eppendorf)

Laboratory shaker

Device for evaporation under a stream of nitrogen

Variably adjustable pipettes (5 to 40, 40 to 200, 200 to 1000 and 1000 to 5000 µL) (e.g. Finn pipettes, from Thermo)

12.5 mL Multipipette (e.g. from Eppendorf)

10 µL and 50 µL Glass syringes (e.g. from Hamilton)

Solid phase extraction station (SPE station) with PTFE valves (e.g. from Baker)

SPE column combination, above 500 mg sulphuric acid/silica gel, below 500 mg benzenesulphonic acid (e.g. J.T. Baker Bakerbond spe™ PCB-A No. 7511-04)

SPE silica gel column (e.g. J.T. Baker Bakerbond spe™ Silica Gel No. 7086-03)

Adapter for combining two SPE columns

Diaphragm pump with manometer and washing bottle

250 mL Separation funnel with Teflon stopcock

10 mL Test tubes with ground glass stoppers (e.g. from Merck)

10 mL Test tubes with Teflon-coated screw caps (e.g. from Merck)

10 mL, 25 mL and 50 mL Volumetric flasks (e.g. from Merck)

25 mL, 100 mL and 250 mL Schott bottles with screw caps and Teflon seals (e.g. from Merck)

100 mL Measuring cylinder (e.g. from Merck)

1.8 mL Crimp-capped vials for the autosampler

100 µL Micro-inserts for the 1.8 mL crimp-capped vials

Glass Pasteur pipettes

2.2 Chemicals

Acetone p.a. (e.g. Merck, No. 1.00014.2500)

Formic acid p.a. (e.g. Merck, No. 1.00264.1000)

n-Hexane for trace analysis (e.g. Scharlau, No. HE0239)

n-Undecane, reference substance for gas chromatography
(e.g. Merck No. 1.09794.0005)

Dichloromethane for trace analysis (e.g. Scharlau No. CL0340)

Bovine blood, defibrinated (e.g. from Elocin lab GmbH)

ADBI (Celestolide®) (e.g. Ehrenstorfer No. LA10045800CY)

AHDI (Phantolide®) (e.g. Ehrenstorfer AHMI No. LA10048000CY)

HHCB (Galaxolide®) (e.g. Ehrenstorfer No. LA14213000CY)

AHTN (Tonalide®) (e.g. Ehrenstorfer No. LA10048500CY)

D_3-AHTN, 100 mg/L in cyclohexane (e.g. Ehrenstorfer No. XA10048600IO)

Helium 4.6 (e.g. from Linde)

Nitrogen 4.6 (e.g. from Linde)

Bidistilled water

2.3 Solutions

Formic acid:
Before use 100 mL formic acid are shaken twice with 25 mL n-hexane in each case in a separation funnel and then stored in a Schott bottle under n-hexane at room temperature. The formic acid is stable for approx. one week when stored under these conditions.

n-Hexane/dichloromethane mixture (1:4, v/v):
20 mL n-hexane and 60 mL dichloromethane are measured separately in a measuring cylinder and mixed in a Schott bottle.
The mixture must be freshly prepared before each analytical series.

2.4 Calibration standards

Starting solutions for the calibration standards:
The solutions of ADBI, AHDI, HHCB and AHTN from Ehrenstorfer (each 10 mg/L) serve as the starting solutions.

Stock solution of the calibration standards (each 200 µg/L):
500 µL of each of the starting solutions are pipetted into a 25 mL volumetric flask. The volumetric flask is then filled to its nominal volume with n-hexane.
This stock solution is also used as a calibration standard solution with a concentration of 200 µg/L. For this purpose the solution is divided into a 10 mL aliquot that is placed in a 10 mL volumetric flask (see pipetting scheme, Table 2).

Working solutions of the calibration standards:
The stock solution (200 µg/L) containing the four musk compounds is diluted as shown in the following pipetting scheme (see Table 2). For this purpose the volumes of the stock solution listed are pipetted into one 10 mL volumetric flask for each calibration point. The volumetric flasks are then filled to their nominal volume with hexane.
The stock and working solutions of the calibration standards thus prepared are filled into labelled Schott bottles and stored at $-18\,°C$. These solutions are stable for at least 1 year when stored under these conditions.

Table 2. Pipetting scheme for the preparation of the calibration standard solutions of the PMCs in hexane

Volume of the stock solution [mL]	Volume of hexane [mL]	Concentration of the calibration standards [µg/L]	Equivalent concentration of the calibration standard in blood [µg/L]*
10	0	200	4.0
5	5	100	2.0
2.5	7.5	50	1.0
1.25	8.75	25	0.5
0.75	9.25	15	0.3
0.5	9.5	10	0.2
0.375	9.625	7.5	0.15
0.25	9.75	5	0.1

* The blood samples are concentrated by 1:50 during processing. The PMC solutions diluted according to this pipetting scheme are prepared in hexane and not processed like the blood samples, therefore the concentration of the hexane solutions must be divided by 50 to obtain the corresponding concentration in blood.

Solution of the internal standard (ISTD):

Starting solution of the ISTD:
The D_3-AHTN solution from Ehrenstorfer (100 mg/L) serves as the starting solution.

Stock solution of the ISTD (10 mg/L):
1 mL of the starting solution of the ISTD is pipetted into a 10 mL volumetric flask. The volumetric flask is then filled to its nominal volume with n-hexane.
When this solution is tightly sealed and stored at $-18\,°C$, it is stable for at least 1 year.

Working solution of the ISTD (1 mg/L):
A pipette is used to transfer 1 mL of the stock solution of the internal standard into a 10 mL volumetric flask. The volumetric flask is then filled to its nominal volume with n-hexane.
When this solution is tightly sealed and stored at $+4$ to $+6\,°C$, it is stable for at least 3 months.

2.5 Conditioning of the SPE columns

An SPE silica gel column is mounted on the connection of the SPE station that is fitted with a PTFE valve. An adapter is used to mount an SPE combination column on the silica gel column. The column combination is then conditioned 4 times, in each case using 2 mL of the hexane/dichloromethane mixture (1:4). After the mixture has completely drained, the column combination is dried by applying a slight vacuum.

3 Specimen collection and sample preparation

The blood is withdrawn in EDTA Monovettes and is either analysed immediately or stored in the deep-freezer at $-18\,°C$. If stored in this way, the blood can be kept for at least 12 months.

3.1 Sample preparation

The problem of contamination of the sample material with PMCs must be taken into consideration in the analytical determination of PMCs in blood samples. The glassware that comes into contact with the biological material in the laboratory must be thoroughly rinsed with acetone and then with hexane. Furthermore, appropriate precautions must be taken at the laboratory workbench (see Section 9.4). Reagent blank values are of great importance for this analysis. For this purpose 2.5 mL bidistilled water is subjected to processing instead of the EDTA blood sample.
Before analysis, the samples are thawed (if necessary) and thoroughly mixed. A pipette is used to transfer 2.5 mL of the EDTA blood sample to a test-tube with a

screw cap. Then 2.5 mL formic acid and 10 µL of the working solution of the ISTD (1 mg/L) are added (beware: foam is formed in the sample, it should be carefully mixed). When the effervescence has stopped, the test-tube is sealed and the mixture is vigorously shaken by hand for approx. 30 s to ensure that a homogeneous phase is obtained. 4 mL hexane are added, the test-tube is sealed again and briefly shaken by hand. Then the test-tube is placed on a laboratory shaker for 30 min to carry out liquid-liquid extraction. The sample is subsequently centrifuged for 5 min at 500 g to separate the phases. A glass Pasteur pipette is used to transfer the organic phase containing the PMCs to a test-tube with a ground glass stopper. A further volume of 4 mL hexane is added to the remaining blood phase, the mixture is shaken by hand as described above, then extraction for 30 min and centrifugation are carried out. The organic phases are combined in the test-tube with the ground glass stopper and the volume is reduced to approx. 0.5 mL by evaporation in a stream of nitrogen.

The extract is purified using the SPE column combination described in Section 2.5. A labelled test-tube with a ground glass stopper is placed in the inner rack of the SPE station to collect the eluate. All the evaporated hexane extract is decanted onto the upper SPE combination column. After the extract has completely seeped in, it is eluted onto the lower silica gel column using 2 mL of the hexane/dichloromethane mixture (1:4) twice (maximum rate 1 drop per second). For this purpose a slight vacuum (up to approx. 600 mbar) can be applied, but it must be ensured that the column does not run dry. The upper column is subsequently dried by applying a slight vacuum, and then it is removed. The lower silica gel column is eluted by introducing 2 mL of the hexane/dichloromethane mixture (1:4) twice without applying a vacuum, and then drained by applying a slight vacuum. 50 µL of n-undecane are added to the eluate as a keeper and the volume of the eluate is reduced to 50 µL by evaporation in a stream of nitrogen. The evaporated extract is transferred to the 100 µL micro-insert, which is placed in a 1.8 mL crimp-capped vial and tightly sealed. The extract is analysed by means of GC-MS/MS. If the instrumental analysis cannot be performed on the same day, the samples should be kept in the refrigerator at approx. +4 to +6 °C and analysed the next day at the latest.

4 Operational parameters

The analytical measurements are performed by a combination of instruments consisting of a gas chromatograph with an autosampler and an ion-trap mass spectrometer with a data processing system.

4.1 Operational parameters for gas chromatography and mass spectrometry

Capillary column: Material: Fused silica
                    Stationary phase: OPTIMA® δ-3
                    Length: 30 m

	Inner diameter:	0.25 mm
	Film thickness:	0.25 μm
Detector:	Ion-trap mass spectrometer	
Temperatures:	Column:	Initial temperature 50 °C, 1 minute isothermal, then increase at a rate of 30 °C/min to 140 °C, then at 5 °C/min to 220 °C, then at 30 °c/min up to 260 °C, 5 min at the final temperature
	Injector:	280 °C
	Transfer line:	280 °C
	Ion-trap:	220 °C
Carrier gas:	Helium 4.6, pre-pressure 12 psi	
Split:	Splitless, split on after 1 min	
Sample volume:	1 μL	
Ionisation type:	Electron impact ionisation (EI)	
MS/MS parameters:	See Table 3	

The measurement conditions listed here were established for the configuration of in-

Table 3. Parameters for the fragmentation of the PMCs in the MS/MS

Substance	Parent ion [m/z]	Mass range [m/z]	Mass width [m/z]	CID [V]	RF amplitude [V]
ADBI/AHDI	229	130–230	3	60	93
HHCB/AHTN/D_3-AHTN	224.5*	140–250	4	58	104

* 224.5 = Mean value of 223 (for non-labelled AHTN) and 226 (for D_3-labelled AHTN); CID: Collision-induced dissociation (non-resonant excitation amplitude); RF: Excitation storage level.

struments used in this case, and they must be optimised for other instruments in accordance with the manufacturer's instructions.

5 Analytical determination

The blood samples processed as described in Section 3.1 are analysed by injecting 1 μL of each n-undecane extract into the gas chromatograph.

In addition to the blood samples, a reagent blank (water instead of blood) and a quality control sample are processed and analysed in each analytical series. If the measured concentrations of the blood samples are not within the linear range of the calibration graph, the blood samples are diluted appropriately and processed anew.

An MS/MS method is used for analytical quantification. Mass spectrometric evaluation is based on daughter ions that are formed in the ion-trap by fragmentation of the parent ions of each PMC (see Table 3 for the list of ions). A mass range is used to evaluate ADBI, AHDI and HHCB, while individual ions of AHTN and D_3-AHTN are evaluated (see Table 4).

Table 4. Retention times and detected mass ranges or individual masses for the evaluation of the PMCs

Analyte	Retention time	Mass range [m/z]	Individual masses [m/z]
ADBI	9 min 15 s	130–230	–
AHDI	9 min 53 s	130–230	–
HHCB	11 min 13 s	140–250	–
AHTN	11 min 22 s	–	187+145
D_3-AHTN	11 min 22 s	–	246+190+146

The retention times shown in Table 4 serve only as a guide. Users of the method must satisfy themselves of the separation power of the capillary column they use and investigate the resulting retention behaviour of the substances.

Figure 2 shows the mass spectra of the PMCs recorded with an ion-trap MS. Figure 3 shows a chromatogram of a PMC calibration standard solution with a concentration of 1 mg/L recorded in the full scan mode (MS) in comparison with a chromatogram recorded in the daughter ion full scan mode (MS/MS). A daughter ion full scan chromatogram (MS/MS) of a spiked bovine blood sample is shown in Figure 4. Figure 5 depicts extracted daughter ion chromatograms for AHTN and D_3-AHTN. Figure 6 shows chromatograms for HHCB after MS/MS analysis of a human blood sample and of a calibration standard solution (4 µg/L hexane).

6 Calibration

A complete calibration graph is plotted for every analytical series.

For this purpose 10 µL of the working solution of the ISTD (1 mg/L) and 50 µL n-undecane are added to 50 µL of the working solution of the calibration standards prepared as described in Section 2.4. The solution is reduced in volume to 50 µL in a stream of nitrogen and analysed as stipulated in Sections 4 and 5.

Calibration graphs are obtained by plotting the quotients of the peak areas of each PMC and that of the internal standard as a function of the concentrations used (see Table 2 for the equivalent concentrations in blood).

The calibration graphs were linear for all the PMCs in the range of 0 to 200 µg/L hexane (equivalent to the concentration range of 0 to 4 µg/L blood).
Figure 7 shows an example of the calibration graphs for each of the four PMCs.

7 Calculation of the analytical result

The concentration in blood is calculated from the calibration graphs (see Section 6). Quotients are calculated by dividing the peak areas of the analytes by that of the internal standard. These quotients are used to read off the pertinent concentration of the PMCs in µg per litre blood from the relevant calibration graph. Appropriate software (e.g. MS Excel) can be used for evaluation. Any reagent blank values must be subtracted from the analytical results for the real samples. If a blood sample has been previously diluted, the analytical result must be multiplied by the dilution factor.

8 Standardisation and quality control

Quality control of the analytical results is carried out as stipulated in the guidelines of the Bundesärztekammer (German Medical Association) [10] and in the special preliminary remarks to this series. In order to determine the precision of the method, a spiked sample of bovine blood containing a constant concentration of PMCs is analysed for each series. As material for quality control is not commercially available, it must be prepared in the laboratory. This quality control sample is prepared by spiking bovine blood with the stock solution of the PMCs (e.g. 100 mL bovine blood with 0.5 mL of the stock solution of the calibration standards (200 µg/L), which results in a concentration of 1 µg/L blood in each case).
The quality control sample is frozen in 3 mL portions. These solutions can be stored for 6 months at $-18\,°C$. The expected value and the tolerance range of this quality control material are determined in a pre-analytical period (one analysis of the control material on each of 20 different days) [11–13].

9 Evaluation of the method

9.1 Precision

Bovine blood was used to determine the precision in the series. It was spiked with 1.0 µg/L of each PMC. This sample was processed and analysed in duplicate on 9 different days. The precision in the series was calculated from the variability of the duplicate determination and the precision from day to day was calculated from the mean values of the duplicate analysis on different days [14, 15]. The results for the precision in the series and the precision from day to day are given in Table 5.

Table 5. Precision in the series and precision from day to day for the determination of the PMCs in bovine blood (n=9)

Substance	Spiked concentration [µg/L]	Precision in the series		Precision from day to day	
		Standard deviation (rel.) [%]	Prognostic range [%]	Standard deviation (rel.) [%]	Prognostic range [%]
ADBI	1.0	14.9	33.7	12.4	28.0
AHDI	1.0	8.7	19.7	16.0	36.2
HHCB	1.0	19.7	44.6	20.0	45.2
AHTN	1.0	14.8	33.5	17.3	39.1

9.2 Accuracy

Recovery experiments were performed to check the accuracy of the method. For this purpose human blood samples without the analytes or spiked with defined quantities of the PMCs were analysed. The spiked concentration of all four PMCs was 1.0 µg/L. The recovery rates for the spiked blood samples are shown in Table 6.

Table 6. Recovery rates for the determination of PMCs in human blood (n=9)

Substance	Spiked concentration [µg/L]	Recovery rate [%]
ADBI	1.0	115
AHDI	1.0	97
HHCB	1.0	115
AHTN	1.0	103

At present no control material is commercially available and no round-robin experiments for the PMCs are offered.

9.3 Detection limits

Under the conditions for sample preparation and gas chromatographic/mass spectrometric determination given here the detection and quantitation limits for PMCs were determined. They were calculated in accordance with DIN 32645 [16]. The detection limit was also estimated as three times the signal/background noise ratio in the temporal vicinity of the analyte signal for a processed blood sample (which did not contain PMCs, however). The results are shown in Table 7.

Table 7. Detection and quantitation limits for the determination of PMCs in blood

Substance	DIN 32 645		3 times the signal/ background noise ratio
	DL [µg/L]	QL [µg/L]	DL [µg/L]
ADBI	0.07	0.27	0.05
AHDI	0.07	0.25	0.1
HHCB	0.09	0.32	0.05
AHTN	0.10	0.38	0.02

DL = Detection Limit, QL = Quantitation Limit.

9.4 Sources of error

Extraneous contamination may occur, even in the laboratory, on account of the ubiquitous occurrence of the PMCs. Therefore it is extremely important to take the following measures to counteract this problem in order to ensure reliable determination of the PMCs:
The workplace used for sample processing should be cleaned prior to analysis without using commercially available disinfectant solutions. They generally contain fragrances that can cause considerable blank values. It is advisable to clean the laboratory workbench with a traditional cleansing agent and then to wipe it several times with clean water. It should be regularly disinfected with a disinfectant solution that is prepared in the laboratory (a 70:30 (v/v) mixture of isopropanol and water). This perfume-free disinfectant solution should also be used at neighbouring workplaces and in communally used laboratory rooms. Furthermore, if laboratory personnel use cream and disinfectant on the hands, these products must be free of perfume.
All the glassware used in the analysis must be stored in a separate place (e.g. in a desiccator) and not with the other glassware, and it must be rinsed with acetone and then with hexane before use.
During solid phase extraction it must be ensured that the elution rate of the combination column is at most one drop per second. If the elution rate is exceeded, the fat is not destroyed to the desired extent. It is carried over and causes interference to the instrumental analysis (recognisable by a drastic deterioration of the signal/background noise ratio).

10 Discussion of the method

The method described here permits the analysis of PMCs in blood by means of GC-MS/MS with very good detection limits and with sufficient precision in the range of less than one microgram per litre.
The signal/background noise ratio is greatly improved (see Figure 3) by the use of MS/MS quantification instead of single MS detection following separation by gas chromatography.

However the HHCB and AHTN peaks are relatively broad in the total ion chromatogram of the daughter ions (see Figure 4). In the case of AHTN this is due to the fact that the retention times of native AHTN and of the internal standard D_3-AHTN are not identical. D_3-AHTN elutes slightly earlier (see Figure 5). Isomeric forms (cis/trans) of HHCB are evidently separated by the OPTIMA® δ-3 column we used [17].

The advantage of the procedure presented here is that only 3 working steps are required: extraction of the blood, purification and reduction in volume of the resulting extracts by evaporation, and then the final measurement. The use of n-undecane as a keeper, which does not evaporate in the stream of nitrogen, allows the measurement volume to be adjusted to 50 µL. The solvent consumption for the entire clean-up procedure is less than 30 mL per sample; the instrumental detection limit for the processed "blank" blood sample is less than 0.1 µg/L. Even at low concentrations the method records a complete mass spectrum of the daughter ion. The specificity is enhanced (see Figure 6) compared with the usual ion chromatograms obtained by a quadrupole mass spectrometer, which normally consist of no more than 3 ion traces. While the method was being checked, a triple quadrupole MS was used for detection instead of an ion-trap MS. The detection limits were one order of magnitude higher in this case. If available, an ion-trap MS should preferably be used for the assay of the PMCs.

A critical point with regard to this analysis is that PMCs are ubiquitous. In particular, cleansing agents (soap, disinfectant solutions, etc.) and skin creams, that are also used in the laboratory, can lead to considerable blank values (see Section 9.4).

The clean-up columns are relatively expensive (approx. € 8 per sample) and ion-trap GC-MS/MS instruments are somewhat rare in medical laboratories.

Alternative clean-up methods that were also investigated were unsuccessful, it proved impossible to separate the PMCs from the fat matrix (Florisil and silica gel with different degrees of activity were tried as well as different eluents). In contrast, the clean-up procedure using sulphuric acid/silica gel, which is also used for analysis of PCBs in sewage sludge and wastewater samples, proved practicable.

Instruments used:
Varian 3400 gas chromatograph, Varian 8200 autosampler, Varian Saturn 3 mass spectrometer (ion-trap with MS/MS option), control and data processing system: Varian Saturn 5.2 software.

11 References

[1] *G. Rimkus:* Polycyclic musk fragrances in the aquatic environment. Toxicol. Lett. 111, 37–56 (1999).
[2] *E.J. van de Plaasche* and *F. Balk:* Environmental risk assessment of the polycyclic musks AHTN and HHCB according to the EU-TGD. RIVM Report No. 601 503 008 (National Institute of Public Health and the Environment (RIVM), Bilthoven, Netherlands (1997).
[3] *H. Gebauer* and *T. Bouter:* Moschus. Euro Cosmetics 1, 30–35 (1997).
[4] *H. Fromme, T. Otto, K. Pilz* and *F. Neugebauer:* Levels of synthetic musks; bromocyclene and PCBs in eel (Anguilla anguilla) and PCBs in sediment samples from waters of Berlin/Germany. Chemosphere 39, 1723–1735 (1999).

[5] *G. Rimkus* and *M. Wolf:* Polycyclic musk fragrances in human adipose tissue and human milk. Chemosphere 33, 2033–2043 (1996).
[6] *H.-D. Eschke, H.-J. Dibowski* and *J. Traud:* Nachweis und Quantifizierung von polycyclischen Moschus-Duftstoffen mittels Ion-Trap GC/MS/MS in Humanfett und Muttermilch. Dtsch. Lebensm.-Rundsch. 91, 375–379 (1995).
[7] *S. Müller, P. Schmid* and *C. Schlatter:* Occurrence of nitro and non-nitro benzenoid musk compounds in human adipose tissue. Chemosphere 33, 17–28 (1996).
[8] *J. Angerer* and *H. U. Käfferlein:* Gas chromatographic method using electron-capture detection for the determination of musk xylene in human blood samples. Biological monitoring of the general population. J. Chrom. B 693, 71–78 (1997).
[9] *K. Bauer* and *C. Frössl:* Blutkonzentrationen von polycyclischen und Nitromoschusverbindungen bei deutschen Probanden. Umwelt-Medizin-Gesellschaft 12, 235–237 (1999).
[10] Bundesärztekammer: Richtlinie der Bundesärztekammer zur Qualitätssicherung quantitativer laboratoriumsmedizinischer Untersuchungen. Dt. Ärztebl. 100, A3335–A3338 (2003).
[11] *J. Angerer* and *G. Lehnert:* Anforderungen an arbeitsmedizinisch-toxikologische Analysen – Stand der Technik. Dt. Ärztebl. 37, C1753–C1760 (1997).
[12] *J. Angerer, Th. Göen* and *G. Lehnert:* Mindestanforderungen an die Qualität von umweltmedizinisch-toxikologischen Analysen. Umweltmed. Forsch. Prax. 3, 307–312 (1998).
[13] *G. Lehnert, J. Angerer* and *K. H. Schaller:* Statusbericht über die externe Qualitätssicherung arbeits- und umweltmedizinisch-toxikologischer Analysen in biologischen Materialien. Arbeitsmed. Sozialmed. Umweltmed. 33, 21–26 (1998).
[14] *W. Funk, V. Dammann* and *G. Donnevert:* Qualitätssicherung in der analytischen Chemie. VCH Verlagsgesellschaft, Weinheim (1992).
[15] *W. Gottwald:* Statistik für Anwender. Wiley-VCH, Weinheim (2000).
[16] DIN 32 645: Nachweis-, Erfassungs- und Bestimmungsgrenze. Beuth Verlag, Berlin (1994).
[17] *S. Biselli, H. Dittmer, R. Gatermann, R. Kallenborn, W.A. König* and *H. Hühnerfuss:* Separation of HHCB, AHTN, ATII and DPMI by enantioselective capillary gas chromatography and preparative separation of HHCB and ATII by enantioselective HPLC. Organohal. Comp. 40, 599–602 (1999).

Authors: *W. Butte, A. Schmidt*
Examiner: *H.-W. Hoppe*

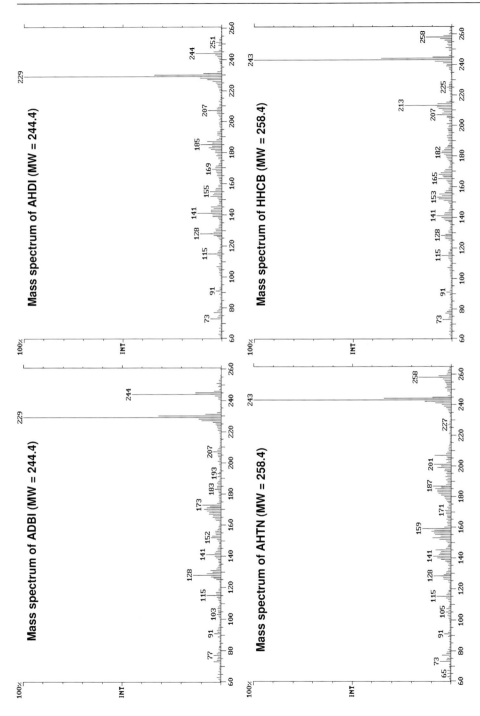

Fig. 2. Mass spectra of the PMCs (recorded by an ion-trap mass spectrometer)

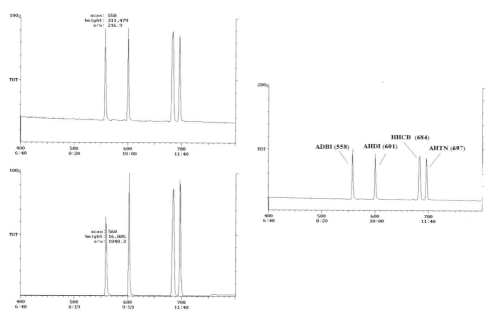

Fig. 3. Comparison of the chromatograms of PMC solutions each at a concentration of 1 mg/L, left above MS (full scan), left below daughter ions (full scan) after MS/MS, right: assignment to the PMCs (scan no.)

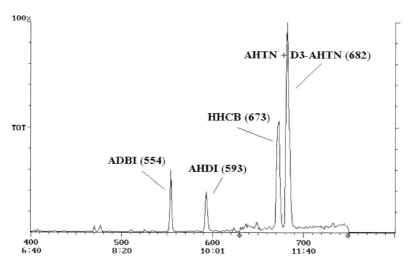

Fig. 4. Daughter ions full scan after MS/MS of a bovine blood sample spiked with 1 μg/L each of ADBI, AHDI, HHCB and AHTN (ISTD=D_3-AHTN, 4 μg/L)

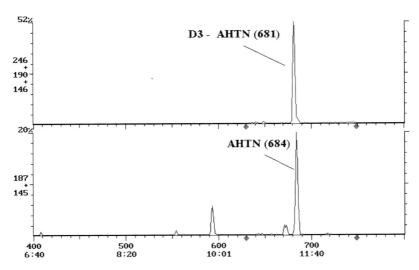

Fig. 5. Extracted daughter ion chromatograms for AHTN and D_3-AHTN

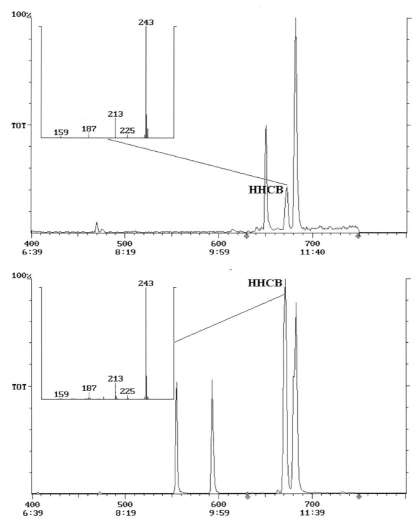

Fig. 6. Chromatograms with daughter ions after MS/MS, above HHCB spectrum in a blood sample (0.12 µg/L), below HHCB spectrum of the calibration standard solution at a concentration of 4 µg/L

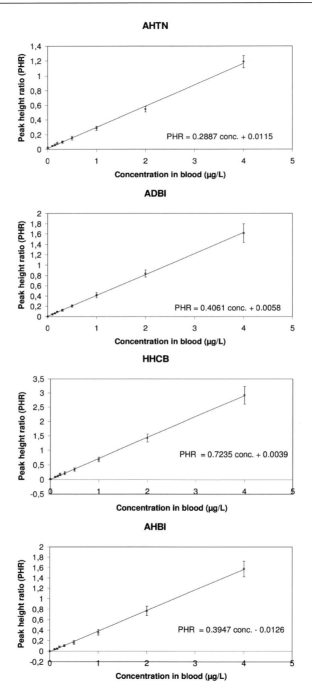

Fig. 7. Calibration graphs for the PMCs

Propylene and diethylene glycol ethers

Application   Determination in blood and urine

Analytical principle   Capillary gas chromatography/
flame ionisation detector (FID)

Summary

This method permits 17 propylene and diethylene glycol ethers at concentrations of relevance to occupational medicine to be determined in blood and urine.

The blood sample to be investigated is diluted with saturated sodium chloride solution or the urine samples are saturated with sodium chloride. Separation of the analytes from the matrix components is carried out by liquid/liquid extraction on diatomaceous earth using a mixture of dichloromethane and acetone for extraction. After a keeper and n-butylbenzene (internal standard) are added, the extract containing the analytes is reduced in volume by evaporation in a stream of nitrogen. The concentrated extract is injected into the gas chromatograph to separate the analytes. Detection of the analytes is carried out by a flame ionisation detector. Calibration is performed using calibration standards prepared in urine or ovine blood and treated in the same manner as the samples to be tested.

Blood

1-Methoxypropanol-2

Within-series imprecision: Standard deviation (rel.) $s_w = 4.5\%$, 8.7% or 5.4%
                             Prognostic range $u = 10.1\%$, 19.5% or 11.9%
                             at a spiked concentration of 2, 5 or 10 mg per litre blood
                             and where n = 10 determinations

Between-day imprecision: Standard deviation (rel.) $s_w = 11.4\%$
                             Prognostic range $u = 26.3\%$
                             at a spiked concentration of 5 mg per litre blood
                             and where n = 8 determinations

Accuracy: Recovery rate $r=80.6\%$ at 2 mg/L
                                                 79.9% at 5 mg/L
                                                 91.1% at 10 mg/L

Detection limit: 0.5 mg 1-methoxypropanol-2 per litre blood

1-Ethoxypropanol-2

Within-series imprecision: Standard deviation (rel.) $s_w=11.5\%$, 5.3% or 4.6%
                          Prognostic range $u=25.5\%$, 11.8% or 10.2%
                          at a spiked concentration of 2, 5 or 10 mg per litre blood
                          and where n = 10 determinations

Between-day imprecision: Standard deviation (rel.) $s_w=14.0\%$
                          Prognostic range $u=32.3\%$
                          at a spiked concentration of 5 mg per litre blood
                          and where n = 8 determinations

Accuracy: Recovery rate $r=81.2\%$ at 2 mg/L
                                                      77.7% at 5 mg/L
                                                      99.2% at 10 mg/L

Detection limit: 1.0 mg 1-ethoxypropanol-2 per litre blood

Diethylene glycol dimethyl ether

Within-series imprecision: Standard deviation (rel.) $s_w=2.6\%$, 2.8% or 2.0%
                          Prognostic range $u=5.8\%$, 6.1% or 4.4%
                          at a spiked concentration of 2, 5 or 10 mg per litre blood
                          and where n = 10 determinations

Accuracy: Recovery rate $r=96.5\%$ at 2 mg/L
                                                      94.0% at 5 mg/L
                                                      97.1% at 10 mg/L

Detection limit: 0.5 mg diethylene glycol dimethyl ether per litre blood

1-Butoxypropanol-2

Within-series imprecision: Standard deviation (rel.) $s_w=3.9\%$, 2.2% or 1.9%
                          Prognostic range $u=8.7\%$, 5.0% or 4.3%
                          at a spiked concentration of 2, 5 or 10 mg per litre blood
                          and where n = 10 determinations

Between-day imprecision: Standard deviation (rel.) $s_w = 7.4\%$
Prognostic range $u = 17.1\%$
at a spiked concentration of 5 mg per litre blood
and where n = 8 determinations

Accuracy: Recovery rate $r = 108.6\%$ at 2 mg/L
97.4% at 5 mg/L
101.6% at 10 mg/L

Detection limit: 0.5 mg 1-butoxypropanol-2 per litre blood

3-Ethoxypropanol-1

Within-series imprecision: Standard deviation (rel.) $s_w = 3.4\%$, 4.1% or 2.8%
Prognostic range $u = 7.5\%$, 9.0% or 6.2%
at a spiked concentration of 2, 5 or 10 mg per litre blood
and where n = 10 determinations

Accuracy: Recovery rate $r = 92.9\%$ at 2 mg/L
85.8% at 5 mg/L
92.1% at 10 mg/L

Detection limit: 1.0 mg 3-ethoxypropanol-1 per litre blood

Diethylene glycol diethyl ether

Within-series imprecision: Standard deviation (rel.) $s_w = 1.9\%$, 2.2% or 1.2%
Prognostic range $u = 4.1\%$, 5.0% or 2.7%
at a spiked concentration of 2, 5 or 10 mg per litre blood
and where n = 10 determinations

Accuracy: Recovery rate $r = 83.6\%$ at 2 mg/L
83.2% at 5 mg/L
85.1% at 10 mg/L

Detection limit: 0.5 mg diethylene glycol diethyl ether per litre blood

Dipropylene glycol monomethyl ether

Within-series imprecision: Standard deviation (rel.) $s_w = 3.4\%$, 2.8% or 1.5%
Prognostic range $u = 7.6\%$, 6.3% or 3.2%
at a spiked concentration of 2, 5 or 10 mg per litre blood
and where n = 10 determinations

Accuracy:	Recovery rate	$r = 112.1\%$ at 2 mg/L
		101.9% at 5 mg/L
		107.8% at 10 mg/L

Detection limit: 1.0 mg dipropylene glycol monomethyl ether per litre blood

Diethylene glycol monomethyl ether

Within-series imprecision: Standard deviation (rel.) $s_w = 5.3\%$, 4.3% or 4.1%
Prognostic range $u = 11.9\%$, 9.5% or 9.1%
at a spiked concentration of 2, 5 or 10 mg per litre blood
and where n = 10 determinations

Accuracy: Recovery rate $r = 67.5\%$ at 2 mg/L
72.2% at 5 mg/L
80.5% at 10 mg/L

Detection limit: 1.0 mg diethylene glycol monomethyl ether per litre blood

Diethylene glycol monoethyl ether

Within-series imprecision: Standard deviation (rel.) $s_w = 6.9\%$, 6.1% or 2.7%
Prognostic range $u = 15.4\%$, 13.6% or 6.1%
at a spiked concentration of 2, 5 or 10 mg per litre blood
and where n = 10 determinations

Accuracy: Recovery rate $r = 91.0\%$ at 2 mg/L
86.5% at 5 mg/L
88.8% at 10 mg/L

Detection limit: 0.5 mg diethylene glycol monoethyl ether per litre blood

Diethylene glycol monobutyl ether

Within-series imprecision: Standard deviation (rel.) $s_w = 3.1\%$, 4.3% or 2.5%
Prognostic range $u = 7.0\%$, 9.5% or 5.7%
at a spiked concentration of 2, 5 or 10 mg per litre blood
and where n = 10 determinations

Accuracy: Recovery rate $r = 109.4\%$ at 2 mg/L
101.6% at 5 mg/L
100.1% at 10 mg/L

Detection limit: 0.5 mg diethylene glycol monobutyl ether per litre blood

Urine

1-Methoxypropanone-2

Within-series imprecision: Standard deviation (rel.) s_w=3.0%, 4.5% or 2.4%
Prognostic range u=6.7%, 10.0% or 5.4%
at a spiked concentration of 2, 5 or 10 mg per litre urine
and where n=10 determinations

Accuracy: Recovery rate r=106.4% at 2 mg/L
103.8% at 5 mg/L
117.4% at 10 mg/L

Detection limit: 0.5 mg 1-methoxypropanone-2 per litre urine

1-Methoxypropanol-2

Within-series imprecision: Standard deviation (rel.) s_w=4.7%, 7.1% or 3.5%
Prognostic range u=10.5%, 15.8% or 7.9%
at a spiked concentration of 2, 5 or 10 mg per litre urine
and where n=10 determinations

Between-day imprecision: Standard deviation (rel.) s_w=7.3%
Prognostic range u=16.8%
at a spiked concentration of 5 mg per litre urine
and where n=8 determinations

Accuracy: Recovery rate r=155.4% at 2 mg/L
125.2% at 5 mg/L
136.2% at 10 mg/L

Detection limit: 1.0 mg 1-methoxypropanol-2 per litre urine

1-Ethoxypropanol-2

Within-series imprecision: Standard deviation (rel.) s_w=4.2%, 5.5% or 4.2%
Prognostic range u=9.4%, 12.3% or 9.4%
at a spiked concentration of 2, 5 or 10 mg per litre urine
and where n=10 determinations

Between-day imprecision: Standard deviation (rel.) s_w=7.4%
Prognostic range u=17.1%
at a spiked concentration of 5 mg per litre urine
and where n=8 determinations

Accuracy:	Recovery rate	$r = 127.1\%$ at 2 mg/L
		103.3% at 5 mg/L
		116.1% at 10 mg/L
Detection limit:	0.5 mg 1-ethoxypropanol-2 per litre urine	

1-Methoxypropyl acetate-2

Within-series imprecision: Standard deviation (rel.) $s_w = 4.2\%$, 3.5% or 2.2%
Prognostic range $u = 9.4\%$, 7.8% or 5.0%
at a spiked concentration of 2, 5 or 10 mg per litre urine and where $n = 10$ determinations

Accuracy:	Recovery rate	$r = 105.4\%$ at 2 mg/L
		99.8% at 5 mg/L
		104.9% at 10 mg/L
Detection limit:	0.5 mg 1-methoxypropyl acetate-2 per litre urine	

1-Ethoxypropyl acetate-2

Within-series imprecision: Standard deviation (rel.) $s_w = 3.1\%$, 2.5% or 2.3%
Prognostic range $u = 6.9\%$, 5.6% or 5.2%
at a spiked concentration of 2, 5 or 10 mg per litre urine and where $n = 10$ determinations

Accuracy:	Recovery rate	$r = 104.9\%$ at 2 mg/L
		100.0% at 5 mg/L
		99.9% at 10 mg/L
Detection limit:	0.5 mg 1-ethoxypropyl acetate-2 per litre urine	

Diethylene glycol dimethyl ether

Within-series imprecision: Standard deviation (rel.) $s_w = 2.6\%$, 2.0% or 2.5%
Prognostic range $u = 5.8\%$, 4.5% or 5.6%
at a spiked concentration of 2, 5 or 10 mg per litre urine and where $n = 10$ determinations

Accuracy:	Recovery rate	$r = 101.1\%$ at 2 mg/L
		100.2% at 5 mg/L
		100.9% at 10 mg/L
Detection limit:	0.5 mg diethylene glycol dimethyl ether per litre urine	

1-Butoxypropanol-2

Within-series imprecision: Standard deviation (rel.) $s_w = 3.5\%$, 1.9% or 5.8%
Prognostic range $u = 7.8\%$, 4.2% or 12.9%
at a spiked concentration of 2, 5 or 10 mg per litre urine
and where n = 10 determinations

Between-day imprecision: Standard deviation (rel.) $s_w = 4.7\%$
Prognostic range $u = 10.8\%$
at a spiked concentration of 5 mg per litre urine
and where n = 8 determinations

Accuracy: Recovery rate $r = 107.8\%$ at 2 mg/L
101.2% at 5 mg/L
98.4% at 10 mg/L

Detection limit: 0.5 mg 1-butoxypropanol-2 per litre urine

3-Ethoxypropanol-1

Within-series imprecision: Standard deviation (rel.) $s_w = 27.7\%$, 17.1% or 8.5%
Prognostic range $u = 61.7\%$, 38.1% or 19.0%
at a spiked concentration of 2, 5 or 10 mg per litre urine
and where n = 10 determinations

Accuracy: Recovery rate $r = 130.2\%$ at 2 mg/L
115.3% at 5 mg/L
109.9% at 10 mg/L

Detection limit: 0.5 mg 3-ethoxypropanol-1 per litre urine

Diethylene glycol diethyl ether

Within-series imprecision: Standard deviation (rel.) $s_w = 2.0\%$, 1.5% or 2.6%
Prognostic range $u = 4.5\%$, 3.3% or 5.9%
at a spiked concentration of 2, 5 or 10 mg per litre urine
and where n = 10 determinations

Accuracy: Recovery rate $r = 104.6\%$ at 2 mg/L
100.7% at 5 mg/L
95.9% at 10 mg/L

Detection limit: 0.5 mg diethylene glycol diethyl ether per litre urine

Propylene glycol diacetate

Within-series imprecision: Standard deviation (rel.) $s_w = 3.8\%$, 1.7% or 2.6%
Prognostic range $u = 8.5\%$, 3.8% or 5.9%
at a spiked concentration of 2, 5 or 10 mg per litre urine
and where n = 10 determinations

Accuracy: Recovery rate $r = 104.0\%$ at 2 mg/L
                                                       99.8% at 5 mg/L
                                                       97.2% at 10 mg/L

Detection limit: 0.5 mg propylene glycol diacetate per litre urine

Dipropylene glycol monomethyl ether

Within-series imprecision: Standard deviation (rel.) $s_w = 21.1\%$, 8.0% or 9.1%
Prognostic range $u = 47.1\%$, 17.8% or 20.3%
at a spiked concentration of 2, 5 or 10 mg per litre urine
and where n = 10 determinations

Accuracy: Recovery rate $r = 109.0\%$ at 2 mg/L
                                                       105.8% at 5 mg/L
                                                       99.1% at 10 mg/L

Detection limit: 2.0 mg dipropylene glycol monomethyl ether per litre urine

Diethylene glycol monomethyl ether

Within-series imprecision: Standard deviation (rel.) $s_w = 11.3\%$, 2.7% or 9.1%
Prognostic range $u = 25.2\%$, 6.0% or 20.3%
at a spiked concentration of 2, 5 or 10 mg per litre urine
and where n = 10 determinations

Accuracy: Recovery rate $r = 112.4\%$ at 2 mg/L
                                                       104.0% at 5 mg/L
                                                       99.2% at 10 mg/L

Detection limit: 1.0 mg diethylene glycol monomethyl ether per litre urine

Diethylene glycol monoethyl ether

Within-series imprecision: Standard deviation (rel.) $s_w = 7.1\%$, 2.9% or 8.7%
Prognostic range $u = 15.8\%$, 6.5% or 19.4%
at a spiked concentration of 2, 5 or 10 mg per litre urine
and where n = 10 determinations

Accuracy:	Recovery rate	$r = 108.5\%$ at 2 mg/L
		101.1% at 5 mg/L
		96.7% at 10 mg/L

Detection limit: 1.0 mg diethylene glycol monoethyl ether per litre urine

Diethylene glycol monoethyl ether acetate

Within-series imprecision: Standard deviation (rel.) $s_w = 3.0\%$, 3.5% or 3.6%
Prognostic range $u = 6.7\%$, 7.8% or 8.1%
at a spiked concentration of 2, 5 or 10 mg per litre urine
and where n = 10 determinations

Accuracy:	Recovery rate	$r = 118.2\%$ at 2 mg/L
		100.0% at 5 mg/L
		97.0% at 10 mg/L

Detection limit: 0.5 mg diethylene glycol monoethyl ether acetate per litre urine

Diethylene glycol dibutyl ether

Within-series imprecision: Standard deviation (rel.) $s_w = 13.7\%$, 7.1% or 65.7%
Prognostic range $u = 30.5\%$, 15.8% or 146.3%
at a spiked concentration of 2, 5 or 10 mg per litre urine
and where n = 10 determinations

Accuracy:	Recovery rate	$r = 104.8\%$ at 2 mg/L
		92.0% at 5 mg/L
		55.0% at 10 mg/L

Detection limit: 1.0 mg diethylene glycol dibutyl ether per litre urine

Diethylene glycol monobutyl ether

Within-series imprecision: Standard deviation (rel.) $s_w = 8.2\%$, 4.1% or 8.7%
Prognostic range $u = 18.3\%$, 9.1% or 19.3%
at a spiked concentration of 2, 5 or 10 mg per litre urine
and where n = 10 determinations

Accuracy:	Recovery rate	$r = 113.4\%$ at 2 mg/L
		103.3% at 5 mg/L
		96.8% at 10 mg/L

Detection limit: 1.0 mg diethylene glycol monobutyl ether per litre urine

Diethylene glycol monobutyl ether actetate

Within-series imprecision: Standard deviation (rel.) $s_w = 5.9\%$, 5.3% or 4.5%
Prognostic range $u = 13.2\%$, 11.8% or 10.0%
at a spiked concentration of 2, 5 or 10 mg per litre urine and where n = 10 determinations

Accuracy: Recovery rate $r = 109.0\%$ at 2 mg/L
99.6% at 5 mg/L
88.2% at 10 mg/L

Detection limit: 0.5 mg diethylene glycol monobutyl ether acetate per litre urine

Propylene and diethylene glycol ethers

The propylene and diethylene glycol ethers that can be determined with this method have quite different physico-chemical properties. Their vapour pressures and boiling points, the viscosity of the liquids and even their odours vary to a large extent, depending on the number of hydroxy groups in the molecules, the presence of an acetyl group and the length of the alkyl chain. The polarity within this class of substances also shows wide variation, ranging from very hydrophilic compounds such as 1-methoxypropanol-2 to notably lipophilic substances diethylene glycol dibutyl ether at the other extreme [1, 2].

From the technical point of view propylene and diethylene glycol ethers are practically indispensable substances for many branches of industry. Their most important application is as solvents and solubilisers in a wide variety of ink and lacquer formulations in the printing industry. Thus, for example, the diethylene glycol methyl ethers and the diethylene glycol ethyl ethers are used in rapid-drying lacquers. Propylene glycol methyl ether and propylene glycol ethyl ether together with their acetates are increasingly employed to delay evaporation or to aid drying. As a result of the steadily growing use of these substances in recent years, their importance has greatly increased from the point of view of occupational medicine [1, 2]. The compounds bearing two ether groups pose more of a problem in the field of environmental medicine on account of their hydrophobic character and their poor biodegradability [2].

The routes of intake of the various propylene and diethylene glycol ethers into the human body vary greatly, depending on the physico-chemical properties of the relevant substance and the type of workplace. For instance, the short-chain glycol ethers 1-methoxypropanol-2, 1-ethoxypropanol-2, diethylene glycol monomethyl ether and the relatively short-chain diethylene glycol dimethyl ether may be present in the air in significant concentrations on account of their rather high vapour pressure, and may therefore be absorbed through the lungs. Transdermal absorption is to be expected in the presence of several of these compounds, e.g. 1-methoxypropanol-2 [1, 2, 3].

Investigations of the metabolism of the substances in the propylene glycol ether group have mainly focused on 1-methoxypropanol-2. It is cleaved at the ether bond

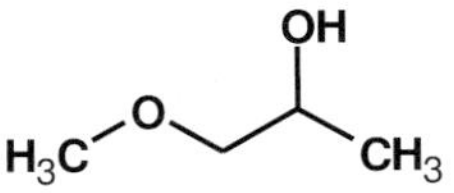

Propylene and diethylene glycol ethers

1-Methoxypropanol-2

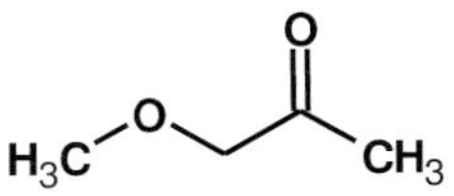

1-Methoxypropanone-2

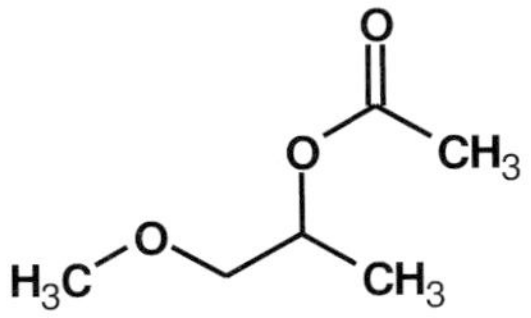

1-Methoxypropyl acetate-2

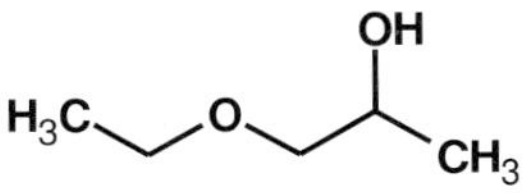

1-Ethoxypropanol-2

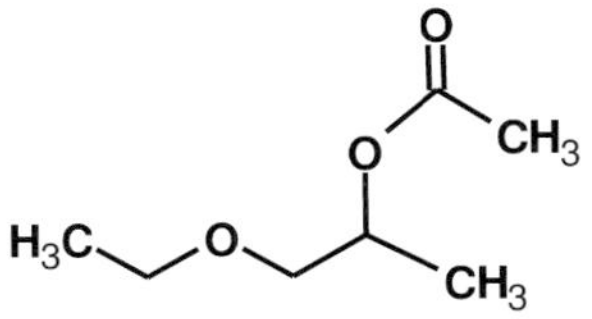

1-Ethoxypropyl acetate-2

3-Ethoxypropanol-1

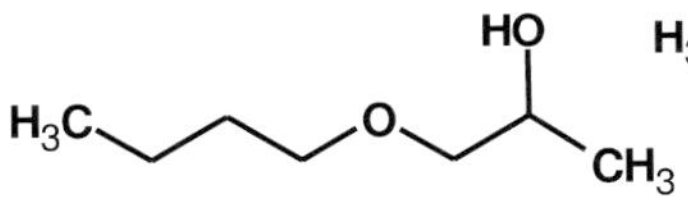

1-Butoxypropanol-2

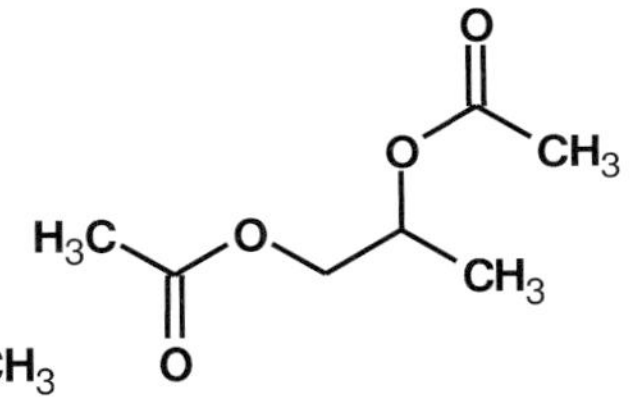

Propylene glycol diacetate

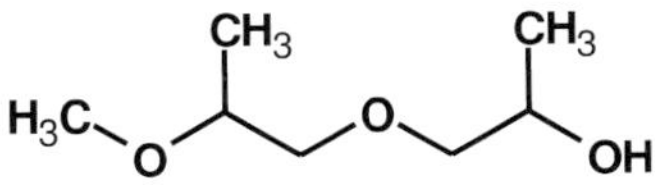

Dipropylene glycol monomethyl ether

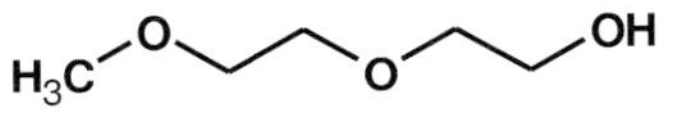

Diethylene glycol monomethyl ether

Diethylene glycol monoethyl ether

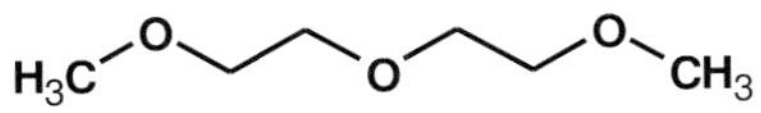

Diethylene glycol dimethyl ether

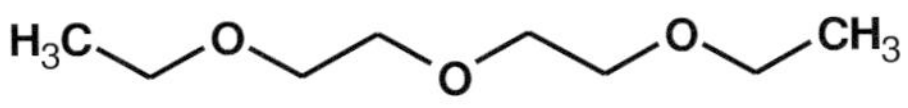

Diethylene glycol diethyl ether

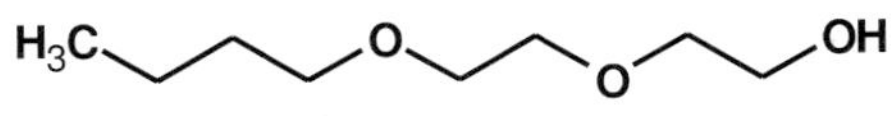

Diethylene glycol monobutyl ether

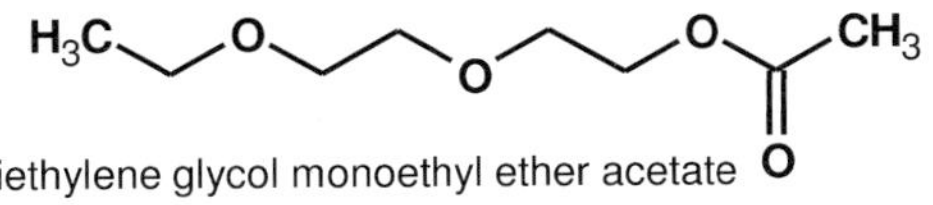

Diethylene glycol monoethyl ether acetate

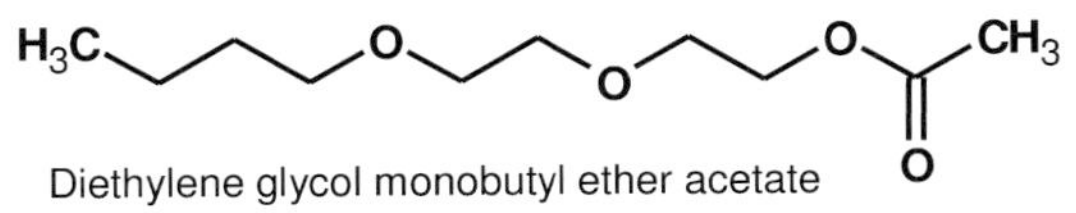

Diethylene glycol monobutyl ether acetate

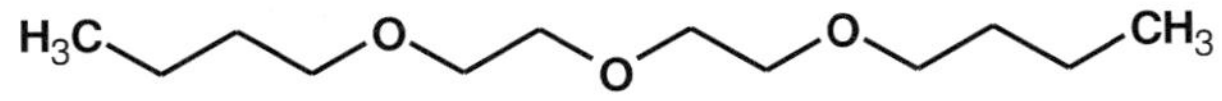

Diethylene glycol dibutyl ether

Table 1. Classification by the MAK Commission [4–11]

Substance	MAK value [mL/m^3]	MAK value [mg/m^3]	Peak limitation category	H[1]; S[2]	Pregnancy risk group
1-Methoxypropanol-2	100	370	I (2)		C
1-Methoxypropyl acetate-2	50	270	I (1)		C
1-Ethoxypropanol-2	50*	220	II (2)	H	C
1-Ethoxypropyl acetate-2	50*	300	II (2)		C
Dipropylene glycol monomethyl ether	50	310	I (1)		D
Diethylene glycol dimethyl ether	5	28	II (8)	H	B
Diethylene glycol monoethyl ether		50 E	I (2)		C
Diethylene glycol monobutyl ether		100	I (1)		C

* MAK value for the sum of the air concentrations of 1-ethoxypropanol-2 and 1-ethoxypropyl acetate-2; 1) Danger of cutaneous absorption, 2) Danger of sensitisation.

Table 2. 1-Methoxypropanol-2 concentrations in the blood and urine of test persons exposed to this substance [2]

Workplace	n =	In urine [mg/L]		In blood [mg/L]	
		Median	Range	Median	Range
Manufacture of hoses	6	3.2	1.7–9.6	14.6	7.3–17.7
Assembly	6	<1.0	<1.0	<1.0	<1.0
Integrity testing	8 (in urine) 7 (in blood)	4.4	1.3–7.2	10.0	7.3–20.1

by enzymes. The resulting alcohols are broken down via several intermediate products to lactic acid and pyruvate, and these compounds are finally metabolised to carbon dioxide and exhaled. A small proportion is excreted by the kidneys in the form of glucuronic acid and sulphuric acid conjugates, as 1,2-propylene glycol, or as the unchanged starting substance. The higher homologues, 1-ethoxypropanol-2 and 1-butoxypropanol-2, are probably metabolised in a similar fashion. Small quantities of the diethylene glycol ethers are also found unchanged in urine together with their cleavage and/or oxidation products. The more lipophilic the substances of this class, the more they are accumulated in the fat-containing compartments of the body. Therefore determination of the higher-molecular diethylene glycol ethers and acetates is preferable in blood rather than in urine [1, 2].

With regard to the toxicological aspects of the propylene and diethylene glycol ethers readers are referred to the relevant justification for the MAK (maximum permitted limit at the workplace) values [4–11]. Table 1 shows the members of the propylene and diethylene glycol ether group that can be analysed with this method and that have been evaluated by the MAK Commission.

There is very little information available on the concentration range of relevance to occupational medicine to be expected in urine and/or in blood. The internal exposure of employees to 1-methoxypropanol-2 was investigated as part of the method validation [2]. The results are shown in Table 2.

Authors: *J. Angerer, Th.Göen, B. Hubner, T. Weiß*
Examiners: *M. Blaszkewicz, B. Aust*

Propylene
and diethylene glycol ethers

Application   Determination in blood and urine

Analytical principle Capillary gas chromatography/
                          flame ionisation detector (FID)

Contents

1 General principles
2 Equipment, chemicals and solutions
2.1 Equipment
2.2 Chemicals
2.3 Solutions
2.4 Calibration standards
2.5 Preparation of the chromatography columns
3 Specimen collection and sample preparation
3.1 Sample preparation
4 Operational parameters for gas chromatography
5 Analytical determination
6 Calibration
7 Calculation of the analytical result
8 Standardisation and quality control
9 Evaluation of the method
9.1 Precision
9.2 Accuracy
9.3 Detection limit
9.4 Sources of error
10 Discussion of the method
11 References

1 General principles

The blood sample to be investigated is diluted with saturated sodium chloride solution or the urine samples are saturated with sodium chloride. Separation of the analytes from the matrix components is carried out by liquid/liquid extraction on diatomaceous earth using a mixture of dichloromethane and acetone for extraction. After

The MAK-Collection Part IV: Biomonitoring Methods, Vol. 11.
DFG, Deutsche Forschungsgemeinschaft
Copyright © 2008 WILEY-VCH Verlag GmbH & Co. KGaA, Weinheim
ISBN: 978-3-527-31596-3

a keeper and n-butylbenzene (internal standard) are added, the extract containing the analytes is evaporated in a stream of nitrogen. The concentrated extract is injected into the gas chromatograph to separate the analytes. Detection of the analytes is carried out by a flame ionisation detector. Calibration is performed using calibration standards prepared in urine or ovine blood and treated in the same manner as the samples to be tested.

2 Equipment, chemicals and solutions

2.1 Equipment

Gas chromatograph with flame ionisation detector (FID) and autosampler as well as an integrator or a PC system for data evaluation.

Capillary gas chromatographic column:
Length: 60 m; inner diameter: 0.32 mm; stationary phase: 100% polyethylene glycol; film thickness: 0.25 µm (e.g. J&W DB-WAX, No. 123-7062)

Disposable syringes containing an anticoagulant (e.g. potassium EDTA Monovettes®, Sarstedt)

10 mL Crimp-capped vials with crimp caps (e.g. from Macherey-Nagel)

250 mL Urine containers (sealable plastic bottles e.g. Sarstedt No. 77.577)

5 µL Syringe for gas chromatography (e.g. Hamilton Microliter™)

Laboratory shaker

Glass chromatography column; length: 30 cm; inner diameter: 15 mm; with frit at the bottom end (e.g. Steiner GmbH No. 867000331)

Holder for serial mounting of 15 to 20 chromatographic columns

Device for evaporation under a stream of nitrogen

Thermostatically controlled water bath

10 mL and 30 mL Polypropylene tubes with stoppers (e.g. from Sarstedt)

10 mL, 50 mL and 100 mL Volumetric flasks

50 µL and 100 µL Pipettes (e.g. from Eppendorf)

Microlitre pipettes adjustable between 100 µL and 1000 µL (e.g. from Eppendorf)

Dispenser fitting for bottles (1 to 10 mL) made of Teflon (e.g. Brand Dispensette®)

1000 mL Measuring cylinder

1 L Screw-capped amber glass bottles

Test-tube shaker (e.g. Vortex, Carl Roth)

1.8 mL Crimp-cap vials with PTFE-coated septa and crimp caps as well as crimping tongs (e.g. from Macherey-Nagel, Düren, Germany)

Micro-inserts for the crimp-cap vials, usable volume 200 μL (e.g. from Macherey-Nagel, Düren, Germany)

2.2 Chemicals

1-Methoxypropanone-2 for synthesis (e.g. VWR No. MOLEM33310079)

1-Methoxypropanol-2, ultrapure (e.g. Merck No. 1.16738.1000)

1-Ethoxypropanol-2, certified (e.g. Ehrenstorfer C 13309000)

1-Methoxypropyl acetate-2 (e.g. VWR No. 52184953)

1-Ethoxypropyl acetate-2 (not commercially available at present)

Diethylene glycol dimethyl ether for synthesis (e.g. Merck No. 8.02934.0250)

1-Butoxypropanol-2 (e.g. TCI No. B0864)

3-Ethoxypropanol-1 (e.g. Fluka No. E7401)

Diethylene glycol diethyl ether for synthesis (e.g. Merck No. 8.02932.0250)

Propylene glycol diacetate for synthesis (e.g. Sigma No. 528072)

Dipropylene glycol monomethyl ether for synthesis (e.g. Merck No. 8.18533.1000)

Diethylene glycol monomethyl ether for synthesis (e.g. Merck No. 8.03128.1000)

Diethylene glycol monoethyl ether for synthesis (e.g. Merck No. 8.03127.1000)

Diethylene glycol monoethyl ether acetate, ultrapure (e.g. Fluka No. 32240)

Diethylene glycol dibutyl ether for synthesis (e.g. Merck No. 8.02933.0250)

Diethylene glycol monobutyl ether for synthesis (e.g. Merck No. 8.03129.1000)

Diethylene glycol monobutyl ether acetate for synthesis (e.g. VWR No. 23833.231)

Sodium chloride p.a. (e.g. Merck No. 1.06404.1000)

Dichloromethane for gas chromatography (e.g. Merck 1.06054.1000)

Acetone for gas chromatography (e.g. Merck No. 1.00012.2500)

Methanol for liquid chromatography (e.g. Merck No. 1.06018.2500)

n-Decane, p.a. (e.g. Fluka No. 30540)

n-Butylbenzene, >99% (e.g. TCI No. B0713)

Diatomaceous earth granulate (e.g. IST Isolute HM-N No. 9800-1000)

EDTA, disodium salt (e.g. VWR No. 443882G)

Nitrogen 5.0 (e.g. from Linde)

Hydrogen 5.0 (e.g. from Linde)

Technical compressed air (e.g. from Linde)

Bidistilled water

2.3 Solutions

Mixture of dichloromethane/acetone 9+1 for extraction:
900 mL dichloromethane and 100 mL acetone are each measured separately in a measuring cylinder and then mixed in a 1 L amber glass bottle. If tightly sealed, this solution can be stored at room temperature for at least 2 months.

Saturated sodium chloride solution:
1 litre bidistilled water is added to 400 g sodium chloride in a 1 L amber glass bottle and the contents are shaken for 15 min. A little sediment of undissolved sodium chloride remains on the bottom of the bottle. This solution is stable at room temperature for at least 1 year.

2.4 Calibration standards

Starting solution:
Approx. 100 mg of each of the various propylene and diethylene glycol ethers are weighed exactly into a 100 mL volumetric flask. The flask is then filled to its nominal volume with methanol (1 g/L).
When stored at $-18\,°C$ the starting solution is stable for at least 1 year.

Stock solution:
A pipette is used to place 10 mL of the starting solution into a 100 mL volumetric flask. The flask is filled to its nominal volume with water for the determination of the propylene and diethylene glycol ethers in blood. For the assay in urine the flask is filled to its nominal volume with pooled urine (100 mg/L).
When stored at $-18\,°C$ the stock solution is stable for at least 2 months.

Calibration standard solutions are prepared in ovine blood using the stock solution (for determination in blood) and in pooled urine (for determination in urine) as shown in the pipetting scheme in Table 3. For this purpose the volumes of the stock solution shown there are pipetted into 100 mL volumetric flasks. The volumetric flasks are subsequently filled to their nominal volume with ovine blood or pooled urine. Using a pipette this material is divided into aliquots of 5 mL (blood) or 8 mL (urine) in sealable 10 mL polypropylene tubes and stored at $-18\,°C$. These solutions are stable for at least 2 months when stored under these conditions.

Table 3. Pipetting scheme for the preparation of the calibration standard solutions

Volume of the stock solution [mL]	Volume of ovine blood or pooled urine* [mL]	Final volume of the calibration standard solution [mL]	Concentration of the calibration standard [mg/L]
2.0	98.0	100	2.0
5.0	95.0	100	5.0
10.0	90.0	100	10.0

* For determination of the propylene and diethylene glycol ethers in blood or urine.

Solution of the internal standard (ISTD)

Stock solution of the ISTD:
100 µL n-butylbenzene are pipetted into a 10 mL volumetric flask. The volumetric flask is then filled to its nominal volume with n-decane.
When stored at $-18\,°C$ the stock solution of the ISTD is stable for at least one year.

Spiking solution of the ISTD:
500 µL of the stock solution of the ISTD are pipetted into a 10 mL volumetric flask. The flask is subsequently filled to its nominal volume with n-decane.
When the spiking solution of the ISTD is tightly sealed and stored at $+4$ to $+6\,°C$, it is stable for at least 2 months.

2.5 Preparation of the chromatography columns

The empty glass chromatography column ($30\,\text{cm} \times 1.5\,\text{cm}$) is filled with 5 g diatomaceous earth granulate. The column material should not be compacted. It is not necessary to condition the column. The sample solutions are introduced directly onto the dry material (see Section 3.1)

3 Specimen collection and sample preparation

Blood

Using a disposable syringe 10 mL of venous blood are withdrawn from the arm vein, and then the contents are gently swirled several times to disperse and dissolve the EDTA anticoagulant.
If analysis is not to be carried out immediately, the blood sample should be transferred to a crimp-capped vial containing approx. 50 mg EDTA (disodium salt), and the vial must be tightly sealed. The samples can be kept for at least 2 months in the deep-freezer at $-18\,°C$. The blood samples are then thawed overnight before analysis.

As soon as the samples have reached room temperature, they must be swirled several times.

Urine

The urine is collected in plastic bottles and should be stored at −18 °C until analysis is performed. Deep-frozen urine can be stored for at least 2 months. The urine samples are allowed to thaw overnight before analysis. As soon as the samples have reached room temperature, they are shaken for 10 min on a laboratory shaker.

3.1 Sample preparation

Blood

In each case 4 mL of the saturated solution of sodium chloride are added to 4 mL of the blood sample in a 10 mL polypropylene tube. Then the tube is sealed and contents are swirled around several times. See below under "Extraction" for the subsequent preparation procedure.

Urine

Approx. 3.5 g sodium chloride are placed in a 10 mL polypropylene tube. Then 8 mL of the urine to be tested are pipetted into the tube and the sample is shaken on a laboratory shaker for approx. 5 minutes. The extraction is carried out as follows:

Extraction

An empty 30 mL PP tube is positioned under the chromatography column prepared as described in Section 2.5 to collect the eluate. The relevant sample solution is then introduced completely onto the chromatography column. After an interval of approx. 20 min (urine) or approx. 30 min (blood) extraction is carried out by adding 5 mL of the extraction mixture (dichloromethane/acetone) 5 times. Approx. 10 to 15 min should elapse between each addition to enable equilibrium to be established between the phases. On completion of the extraction, i.e. when no more extract emerges from the chromatography column, 50 µL of the solution of the ISTD are added to the eluate. Then the contents of the tube are thoroughly mixed (Vortex). The sample is finally evaporated from a volume of approx. 25 mL to about 100 to 200 µL in a water bath at 25 °C in a stream of nitrogen. The concentrated extract is finally transferred to a 1.8 mL crimp-capped vial with a micro-insert, and the vial is sealed. 4 µL of this solution is injected into the gas chromatograph for analysis.

4 Operational parameters for gas chromatography

Capillary column:	Material:	Fused silica
	Stationary phase:	DB-WAX
	Length:	60 m
	Inner diameter:	0.32 mm
	Film thickness:	0.25 µm
Detector:	Flame ionisation detector (FID)	
Temperatures:	Column:	Initial temperature 65 °C, 10 min isothermal, then increase at a rate of 10 °C/min to 180 °C, 6 min isothermal, then increase at 50 °C/min to 230 °C, 2 min at the final temperature
	Injector:	230 °C
	Detector:	250 °C
Carrier gas:	Nitrogen 5.0 with a column pre-pressure of 1034 hPa (15 psi)	
Combustion gases:	Compressed air at a flow rate of 300 mL/min Hydrogen, flow rate 30 mL/min	
Split:	15 mL/min	
Injection volume:	4 µL	

All other parameters must be optimised in accordance with the manufacturer's instructions.

5 Analytical determination

The urine samples or blood samples processed as described in Section 3.1 are analysed by injecting 4 µL of each extract into the gas chromatograph. If the measured values are above the range of the calibration graph (>10 mg/L), the samples are appropriately diluted with water and processed anew.

A quality control sample is analysed with each analytical series. Figures 1 and 2 each show a chromatogram of a spiked blood and a spiked urine sample. The retention times shown there serve only as guidelines. Users of the method must satisfy themselves of the separation power of the capillary column used and the resulting retention behaviour of the substances.

6 Calibration

The calibration standard solutions in urine or in ovine whole blood are processed in the same manner as the native samples as described in Section 3.1 and analysed by gas chromatography using a flame ionisation detector as stipulated in Sections 4 and 5. Linear calibration graphs are obtained by plotting the quotients of the peak areas of the analyte and that of the internal standard as a function of the concentrations used. It is not necessary to plot a complete calibration graph for every analytical series. It is sufficient to include one calibration standard in the analysis each day. The ratio of the result obtained for this standard and the result for the equivalent standard in the complete calibration graph is calculated. Using this quotient, each result read off the calibration graph is corrected (one-point calibration).

New calibration graphs should be plotted if the quality control results indicate systematic deviation.

The calibration graph is linear between the detection limit and 10 mg per litre urine or blood.

7 Calculation of the analytical result

Quotients are calculated by dividing the peak areas of the analytes by that of the internal standard. These quotients are used to read off the pertinent concentration of the analytes in mg per litre urine or blood from the relevant calibration graph. If the pooled urine used to prepare the calibration standards exhibits a background signal, the resulting calibration graph must be shifted in parallel so that it passes through the zero point of the coordinates. The concentrations of the background exposure can be read off from the point where the graph intercepts the axis before parallel shifting in each case. As a rule, the ovine blood used to prepare the calibration standard solutions is free from background interference. No reagent blank value has been ascertained during analysis to date.

8 Standardisation and quality control

Quality control of the analytical results is carried out as stipulated in the guidelines of the Bundesärztekammer (German Medical Association) [12] and in the special preliminary remarks to this series. In order to determine the precision of the method a urine sample or a blood sample containing a constant concentration of analytes is tested. As material for quality control is not commercially available, it must be prepared in the laboratory. For this purpose pooled urine or commercially available ovine blood is spiked with a defined quantity of the analytes. The concentration of this control material should lie within the relevant concentration range. A six-month supply of the control material is prepared, divided into aliquots in 10 mL polypropylene tubes and stored in the deep-freezer. When stored under these conditions, it can be used for 6 months. The expected value and the tolerance range of this quality control material are determined in a pre-analytical period (one analysis of the control material on each of 20 different days) [13–15].

9 Evaluation of the method

9.1 Precision

Blood

Ovine blood with no exposure to the analytes was spiked with 2 mg/L, 5 mg/L or 10 mg/L of the analytes and then processed and analysed to determine the precision in the series. Ten replicate determinations of each of the blood samples yielded the precision in the series shown in Table 4.

Table 4. Precision in the series in ovine blood (n = 10)

Substance	Mean value [mg/L]	Standard deviation (rel.) [%]	Prognostic range [%]
1-Methoxypropanol-2	1.44 3.56 8.13	4.5 8.7 5.4	10.1 19.5 11.9
1-Ethoxypropanol-2	1.40 3.36 8.57	11.5 5.3 4.6	25.5 11.8 10.2
Diethylene glycol dimethyl ether	1.81 4.40 9.09	2.6 2.8 2.0	5.8 6.1 4.4
1-Butoxypropanol-2	1.82 4.07 8.49	3.9 2.2 1.9	8.7 5.0 4.3
3-Ethoxypropanol-1	1.55 3.76 8.07	3.4 4.1 2.8	7.5 9.0 6.2
Diethylene glycol diethyl ether	1.46 3.63 7.42	1.9 2.2 1.2	4.1 5.0 2.7
Dipropylene glycol monomethyl ether	2.03 4.62 9.77	3.4 2.8 1.5	7.6 6.3 3.2
Diethylene glycol monomethyl ether	1.39 3.70 8.26	5.3 4.3 4.1	11.9 9.5 9.1
Diethylene glycol monoethyl ether	1.74 4.08 8.38	6.9 6.1 2.7	15.4 13.6 6.1
Diethylene glycol monobutyl ether	1.98 4.59 9.05	3.1 4.3 2.5	7.0 9.5 5.7

Urine

Pooled urine with no background exposure to the analytes was spiked with 2 mg/L, 5 mg/L or 10 mg/L of the analytes and then processed and analysed to determine the precision in the series. Ten replicate determinations of each of the urine samples yielded the precision in the series shown in Table 5.

Table 5. Precision in the series in urine (n=10)

Substance	Mean value [mg/L]	Standard deviation (rel.) [%]	Prognostic range [%]
1-Methoxypropanone-2	2.00	3.0	6.7
	4.33	4.5	10.0
	11.10	2.4	5.4
1-Methoxypropanol-2	2.58	4.7	10.5
	5.47	7.1	15.8
	11.91	3.5	7.9
1-Ethoxypropanol-2	2.16	4.2	9.4
	4.40	5.5	12.3
	9.92	4.2	9.4
1-Methoxypropyl acetate-2	1.95	4.2	9.4
	4.61	3.5	7.8
	9.73	2.2	5.0
1-Ethoxypropyl acetate-2	1.93	3.1	6.9
	4.60	2.5	5.6
	9.21	2.3	5.2
Diethylene glycol dimethyl ether	1.80	2.6	5.8
	4.47	2.0	4.5
	9.02	2.5	5.6
1-Butoxypropanol-2	1.80	3.5	7.8
	4.22	1.9	4.2
	8.24	5.8	12.9
3-Ethoxypropanol-1	2.24	27.7	61.7
	4.97	17.1	38.1
	9.52	8.5	19.0
Diethylene glycol diethyl ether	1.82	2.0	4.5
	4.39	1.5	3.3
	8.41	2.6	5.9
Propylene glycol diacetate	2.10	3.8	8.5
	5.04	1.7	3.8
	9.84	2.6	5.9
Dipropylene glycol monomethyl ether	1.93	21.1	47.1
	4.68	8.0	17.8
	8.81	9.1	20.3

Table 5 (continued)

Substance	Mean value [mg/L]	Standard deviation (rel.) [%]	Prognostic range [%]
Diethylene glycol monomethyl ether	2.27 4.96 9.54	11.3 2.7 9.1	25.2 6.0 20.3
Diethylene glycol monoethyl ether	2.04 4.73 9.11	7.1 2.9 8.7	15.8 6.5 19.4
Diethylene glycol monoethyl ether acetate	2.27 4.81 9.33	3.0 3.5 3.6	6.7 7.8 8.1
Diethylene glycol dibutyl ether	1.74 3.81 4.61	13.7 7.1 65.7	30.5 15.8 146.3
Diethylene glycol monobutyl ether	2.03 4.63 8.75	8.2 4.1 8.7	18.3 9.1 19.3
Diethylene glycol monobutyl ether acetate	2.05 4.70 8.32	5.9 5.3 4.5	13.2 11.8 10.0

Furthermore, the precision from day to day was also determined for 1-methoxypropanol-2, 1-ethoxypropanol-2 and 1-butoxypropanol-2. For this purpose pooled urine and ovine whole blood with no background exposure to these substances were spiked with 5 mg/L of the above-mentioned analytes and then processed and analysed on 8 different days. The resulting precision from day to day is shown in Table 6.

Table 6. Precision from day to day in blood and urine (n=8)

Substance	Mean value [mg/L]	Standard deviation (rel.) [%]	Prognostic range [%]
Blood			
1-Methoxypropanol-2	5.35	11.4	26.3
1-Ethoxypropanol-2	5.27	14.0	32.3
1-Butoxypropanol-2	5.02	7.4	17.1
Urine			
1-Methoxypropanol-2	5.14	7.3	16.8
1-Ethoxypropanol-2	5.06	7.4	17.1
1-Butoxypropanol-2	4.94	4.7	10.8

9.2 Accuracy

Recovery experiments were performed to check the accuracy of the method. The same material that was used to determine the precision in the series was analysed, and this material was processed and analysed 10 times as described in the previous sections. Evaluation was carried out with the aid of calibration standard solutions that were prepared in water. The relative recovery rates for the individual propylene and diethylene glycol ethers for the matrices blood and urine at three different concentrations are shown in Table 7.

Table 7. Relative recovery rates for the propylene and diethylene glycol ethers in blood and in urine

Substance	Rel. recovery rate [%] in					
	Blood			Urine		
Spiked with:	2 mg/L	5 mg/L	10 mg/L	2 mg/L	5 mg/L	10 mg/L
1-Methoxypropanone-2	n.d.	n.d.	n.d.	106.4	103.8	117.4
1-Methoxypropanol-2	80.6	79.9	91.1	155.4	125.2	136.2
1-Ethoxypropanol-2	81.2	77.7	99.2	127.1	103.3	116.1
1-Methoxypropyl acetate-2	n.d.	n.d.	n.d.	105.4	99.8	104.9
1-Ethoxypropyl acetate-2	n.d.	n.d.	n.d.	104.9	100.0	99.9
Diethylene glycol dimethyl ether	96.5	94.0	97.1	101.1	100.2	100.9
1-Butoxypropanol-2	108.6	97.4	101.6	107.8	101.2	98.4
3-Ethoxypropanol-1	92.9	85.8	92.1	130.2	115.3	109.9
Diethylene glycol diethyl ether	83.6	83.2	85.1	104.6	100.7	95.9
Propylene glycol diacetate	n.d.	n.d.	n.d.	104.0	99.8	97.2
Dipropylene glycol monomethyl ether	112.1	101.9	107.8	109.0	105.8	99.1
Diethylene glycol monomethyl ether	67.5	72.2	80.5	112.4	104.0	99.2
Diethylene glycol monoethyl ether	91.0	86.5	88.8	108.5	101.1	96.7
Diethylene glycol monoethyl ether acetate	n.d.	n.d.	n.d.	118.2	100.0	97.0
Diethylene glycol dibutyl ether	n.d.	n.d.	n.d.	104.8	92.0	55.0
Diethylene glycol monobutyl ether	109.4	101.6	100.1	113.4	103.3	96.8
Diethylene glycol monobutyl ether acetate	n.d.	n.d.	n.d.	109.0	99.6	88.2

n.d. = not determined.

9.3 Detection limit

The detection limit of the various glycol ethers ranges between 0.5 mg/L and 2.0 mg/L under the conditions given here for sample preparation and determination by means of GC-FID. As no reagent blank values occurred, a signal/background noise ratio of 3 was used to establish the detection limits. Table 8 shows the detection limits of the analytes for the matrices blood and urine.

Table 8. Detection limits for the propylene and diethylene glycol ethers in blood and in urine

Substance	Blood [mg/L]	Urine [mg/L]
1-Methoxypropanone-2	n.d.	0.5
1-Methoxypropanol-2	0.5	1.0
1-Ethoxypropanol-2	1.0	0.5
1-Methoxypropyl acetate-2	n.d.	0.5
1-Ethoxypropyl acetate-2	n.d.	0.5
Diethylene glycol dimethyl ether	0.5	0.5
1-Butoxypropanol-2	0.5	0.5
3-Ethoxypropanol-1	1.0	0.5
Diethylene glycol diethyl ether	0.5	0.5
Propylene glycol diacetate	n.d.	0.5
Dipropylene glycol monomethyl ether	1.0	2.0
Diethylene glycol monomethyl ether	1.0	1.0
Diethylene glycol monoethyl ether	0.5	1.0
Diethylene glycol monoethyl ether acetate	n.d.	0.5
Diethylene glycol dibutyl ether	n.d.	1.0
Diethylene glycol monobutyl ether	0.5	1.0
Diethylene glycol monobutyl ether acetate	n.d.	0.5

n.d. = not determined.

9.4 Sources of error

One of the most important steps in this method is the liquid/liquid extraction of the analytes on the diatomaceous earth material. This material has a very high inner surface area and adsorbs the entire sample solution. The extraction is performed using a mixture of dichloromethane and acetone. It is important in this context that the sample volume introduced for analysis and the quantity of diatomaceous earth are in the optimum ratio to each other. During method validation a glass chromatography column with dimensions of 30 cm × 1.5 cm filled with 5 g of the diatomaceous earth in combination with a sample volume of 8 mL proved to be optimal. After the sample has been completely transferred, about 20% of the diatomaceous earth layer at the end of the column should still remain unwetted. This "clean-up zone" has two functions. Firstly, the dichloromethane/acetone crude extract that is saturated with water is dried in the clean-up zone, making it unnecessary to add drying agents such as sodium sulphate etc. before gas chromatographic analysis. Secondly, various additional co-extracted interfering substances in the urine, such as urea, uric acid, creatinine, organic acids etc., are retained in this zone.

The given water bath temperature of 25 °C should not be exceeded when the eluate from the diatomaceous earth column is reduced in volume by evaporation. In addition to the reference substance n-butylbenzene, n-decane is added to the solution of the ISTD. This functions as a keeper while the eluate is being concentrated in the stream of nitrogen and prevents loss of the analytes.

Close attention must be paid to the condition of the separation column, which is especially decisive for the chromatographic separation and the shape of the peaks of the relatively polar analytes. If deterioration in the separation power is observed, it is advisable to shorten the start of the column by 0.5 to 1.0 metre.

10 Discussion of the method

The method described here permits 17 different propylene and diethylene glycol ethers to be determined in one single analytical run. The propylene glycol ethers 1-propoxypropanol-2 and 1-phenoxypropanol-2 were not taken into consideration during method development. However, it can be assumed that this procedure is also capable of determining these substances with sufficient precision and accuracy.

The extraction procedure used in this case is considerably more practical, rapid and efficient in every way than multiple liquid/liquid extractions, which are based on a similar principle. The extraction yield, especially of the short-chain compounds, was enhanced by the use of dichloromethane with a 10% admixture of acetone instead of pure dichloromethane. The addition of sodium chloride to saturate the aqueous sample led to a dramatic increase in the yield of the short-chain and therefore usually the more hydrophilic compounds.

In order to achieve the desired sensitivity at all, the extract must be concentrated to about one hundredth of its original volume. Preliminary experiments showed that compliance with the stipulated water bath temperature of 25 °C and the addition of n-decane as a keeper prevent losses due to evaporation during this step.

Moreover, it was shown that n-butylbenzene has proved a good choice as an internal standard for routine application of the method. No notable evaporation losses are to be expected due to its low vapour pressure. The most important step of the procedure, the extraction of the analytes from the aqueous sample matrix, proceeds to almost exhaustion, as the volume of the eluent is sufficiently large. For this reason internal standardisation of this step proved unnecessary. On principle, however, it is possible to add the internal standard at the beginning of the sample preparation. But in this case the fact that n-butylbenzene would only be suitable as an internal standard for some of the analytes during this step must be taken into account, as the analytes show differences in their extraction behaviour. Therefore it is impossible to find a universal internal standard for the whole range from strongly polar to non-polar substances during the extraction process.

Figures 1 and 2 show the chromatograms of a calibration standard solution in blood and urine that each contain the various glycol ethers in concentrations of 5 and 10 mg/L respectively. It is evident that the polyethylene glycol phase of the capillary column permits baseline separation of the 17 different substances. Furthermore, any interfering substances that may be present in the blood or urine are largely eliminated. And finally, the wide range of different substances that can be determined by this method in the case of exposure to a mixture of several of the test substances is of great advantage.

The precision of the method described here for the assay of the propylene and diethylene glycol ethers in blood is regarded as good for the majority of the tested substances. The precision of the determination of the two short-chain glycol ethers 1-methoxypropanol-2 and 1-ethoxypropanol-2 is somewhat poorer than that achieved for the other glycol ethers. The precision data for the assay in urine, with a relative standard deviation of between 2 and 11%, are less favourable than the values obtained for blood on account of the matrix-dependent analytical background interference. In particular, no satisfactory results were obtained for certain compounds, such as 3-ethoxypropanol-1, dipropylene glycol monomethyl ether and diethylene glycol dibutyl ether, with relative standard deviations in the range of approx. 17 to 65%. The reason for these poor results is largely unknown.

The mean recovery rates between 68 and 112% for the glycol ethers in blood are not as satisfactory as the values for the determination in urine, for which values of 95 to 110% were achieved. In contrast, 1-methoxypropanol-2 (125 to 155%) and diethylene glycol dibutyl ether (55 to 105%) yielded poorer results. In the case of the former substance, peak resolution and peak integration were not optimal due to a contaminated capillary column. A considerable improvement in the separation power of the column was achieved by shortening the column at the contaminated injector end by a length of about 0.5 to 1.0 m. In the case of the latter, very hydrophobic substance the losses are probably a consequence of solubility problems when the starting solution in methanol is diluted with urine. If this substance is to be assayed, it is advisable to prepare a separate stock solution for it. This is then further diluted with methanol instead of urine in the first dilution step. The appropriate amount of this solution is then added to the actual calibration standard solution.

The detection limits of the individual substances were calculated as three times the background noise at the relevant retention time of each analyte. The values are between 0.5 and 1.0 mg/L for the individual glycol ethers in blood and between 0.5 and 2.0 mg/L in urine. This seems adequate in view of the internal exposures that occur in biomonitoring in the field of occupational medicine, which usually clearly exceed the detection limits of the various glycol ethers. The method is therefore very suitable for this application.

The practicability of this procedure can be regarded as favourable. However, a dispenser fitting should be used to introduce the eluent onto the columns in every case in order to save time.

If, in addition, a holder for up to 20 chromatography columns is constructed in the laboratory and used in this case, it is possible for a trained laboratory employee to prepare 20 samples for subsequent GC analysis by the end of one working day.

Instruments used:
Varian 3400 gas chromatograph with a flame ionisation detector, AS 8100 autosampler and Varian data system workstation.

11 References

[1] *E. Bingham, B. Cohrssen* and *C.H. Powell (eds.):* Patty's Toxicology. 5th Edition. Volume VII. John Wiley & Sons, New York (2001).

[2] *B. Hubner:* Analytische Verfahren zur Bestimmung von Propylenglykol, Diethylenglykol, ihren Alkylethern und den Stoffwechselprodukten dieser Verbindungen in Körperflüssigkeiten zur Abschätzung eines berufsbedingten Gesundheitsrisikos. Universität Erlangen-Nürnberg. Naturwiss. Fakultät, Diss. (1993).

[3] *B. Hubner, G. Lehnert, K.H. Schaller, D. Welte* and *J. Angerer:* Chronic occupational exposure to organic solvents. XV. Glycol ether exposure during the manufacture of brakehoses. Int. Arch. Occup. Environ. Health 64(4), 261–264 (1992).

[4] *Deutsche Forschungsgemeinschaft:* MAK- und BAT-Werte-Liste 2006, 42nd issue, Wiley-VCH, Weinheim (2006).

[5] *H. Greim (ed.):* 1-Methoxy-2-propanol. Occupational Toxicants – Critical data evaluation for MAK values and classification of carcinogens. Vols. 5 and 14. Wiley-VCH, Weinheim (1993, 2000).

[6] *H. Greim (ed.):* 1-Methoxypropylacetat-2. Gesundheitsschädliche Arbeitsstoffe. Toxikologisch-arbeitsmedizinische Begründungen von MAK-Werten. 17th and 30th issues. Wiley-VCH, Weinheim (1991, 2000).

[7] *H. Greim (ed.):* 1-Ethoxy-2-propanol. Gesundheitsschädliche Arbeitsstoffe. Toxikologisch-arbeitsmedizinische Begründungen von MAK-Werten. 43rd issue. Wiley-VCH, Weinheim (2007).

[8] *H. Greim (ed.):* 1-Ethoxy-2-propylacetat. Gesundheitsschädliche Arbeitsstoffe. Toxikologisch-arbeitsmedizinische Begründungen von MAK-Werten. 42nd issue. Wiley-VCH, Weinheim (2007).

[9] *H. Greim (ed.):* Dipropylenglykolmonomethylether. Gesundheitsschädliche Arbeitsstoffe. Toxikologisch-arbeitsmedizinische Begründungen von MAK-Werten. 12th and 30th issues. Wiley-VCH, Weinheim (1986, 2000).

[10] *H. Greim (ed.):* Diethylenglykoldimethylether. Gesundheitsschädliche Arbeitsstoffe. Toxikologisch-arbeitsmedizinische Begründungen von MAK-Werten. 20th and 33rd issues. Wiley-VCH, Weinheim (1994, 2001).

[11] *H. Greim (ed.):* Butyldiglykol. Gesundheitsschädliche Arbeitsstoffe. Toxikologisch-arbeitsmedizinische Begründungen von MAK-Werten. 18th and 30th issues. Wiley-VCH, Weinheim (1992, 2000).

[12] *Bundesärztekammer:* Richtlinie der Bundesärztekammer zur Qualitätssicherung quantitativer laboratoriumsmedizinischer Untersuchungen. Dt. Ärztebl. 100, A3335–A3338 (2003).

[13] *J. Angerer* and *G. Lehnert:* Anforderungen an arbeitsmedizinisch-toxikologische Analysen – Stand der Technik. Dt. Ärztebl. 37, C1753–C1760 (1997).

[14] *J. Angerer, Th. Göen* and *G. Lehnert:* Mindestanforderungen an die Qualität von umweltmedizinisch-toxikologischen Analysen. Umweltmed. Forsch. Prax. 3, 307–312 (1998).

[15] *G. Lehnert, J. Angerer* and *K.H. Schaller:* Statusbericht über die externe Qualitätssicherung arbeits- und umweltmedizinisch-toxikologischer Analysen in biologischen Materialien. Arbeitsmed. Sozialmed. Umweltmed. 33(1), 21–26 (1998).

Authors: *J. Angerer, Th. Göen, B. Hubner, T. Weiß*
Examiners: *M. Blaszkewicz, B. Aust*

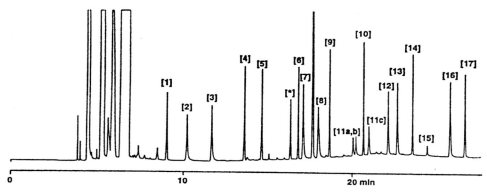

Fig. 1. Chromatogram of a worked-up urine sample spiked with 5 mg/L of the analytes.
1 1-Methoxypropanone-2; **2** 1-Methoxypropanol-2; **3** 1-Ethoxypropanol-2; **4** 1-Methoxypropyl acetate-2; **5** 1-Ethoxypropyl acetate-2; **6** Diethylene glycol dimethyl ether; **7** 1-Butoxypropanol-2; **8** 3-Ethoxypropanol-1; **9** Diethylene glycol diethyl ether, **10** Propylene glycol diacetate; **11** Dipropylene glycol monomethyl ether; **12** Diethylene glycol monomethyl ether; **13** Diethylene glycol monoethyl ether, **14** Diethylene glycol monoethyl ether acetate; **15** Diethylene glycol dibutyl ether; **16** Diethylene glycol monobutyl ether; **17** Diethylene glycol monobutyl ether acetate; * Internal standard n-butylbenzene

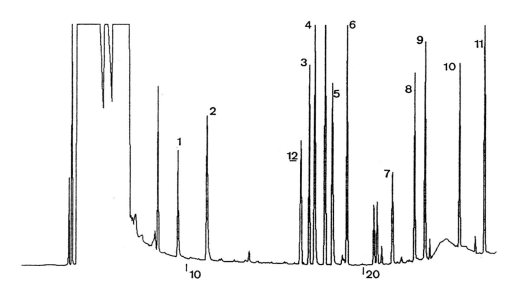

Fig. 2. Chromatogram of a worked-up blood sample spiked with 10 mg/L of the analytes.
1 1-Methoxypropanol-2; **2** 1-Ethoxypropanol-2; **3** Diethylene glycol dimethyl ether; **4** 1-Butoxypropanol-2; **5** 3-Ethoxypropanol-1; **6** Diethylene glycol diethyl ether; **7** Dipropylene glycol monomethyl ether, **8** Diethylene glycol monomethyl ether; **9** Diethylene glycol monoethyl ether; **10** Diethylene glycol dibutyl ether; **11** Diethylene glycol monobutyl ether; **12** Internal standard n-butylbenzene

2,3,7,8-Tetrachlorodibenzo-p-dioxin (TCDD)

Application   Determination in blood

Analytical principle   Capillary gas chromatography/
mass spectrometric detection (MS)

Summary

The method described here permits the determination of 2,3,7,8-tetrachlorodibenzo-p-dioxin (2,3,7,8-TCDD) in blood. Exposure to 2,3,7,8-TCDD of relevance to occupational medicine can thus be reliably measured.

For this purpose the blood is initially subjected to liquid-liquid extraction. The analyte is subsequently enriched on an aluminium oxide solid phase and separated from matrix components. After separation by capillary gas chromatography, the analyte is measured by means of high-resolution mass spectrometry and electron impact ionisation (EI). Linear calibration graphs are used for quantitative evaluation. The standard solutions required for calibration are prepared with pooled human blood to which defined quantities of the calibration standard are added. The standard solutions are treated in the same manner as the blood samples to be investigated. ^{13}C-2,3,7,8-TCDD is added to the blood samples as an internal standard.

2,3,7,8-Tetrachlorodibenzo-p-dioxin

Within-series imprecision: Standard deviation (rel.) $s_w = 9.1\%$ or 10.4%
Prognostic range $u = 22.3\%$ or 24.4%
at spiked concentrations of 30 or 100 pg per litre blood
and where n = 6 determinations

Between-day imprecision: Standard deviation (rel.) $s_w = 7.0\%$
Prognostic range $u = 17.3\%$
at a spiked concentration of 100 pg per litre blood
and where n = 6 determinations

Accuracy: Recovery rate $r = 106\%$ at 30 pg/L
and   108% at 100 pg/L

Detection limit: 6 pg 2,3,7,8-tetrachlorodibenzo-p-dioxin per litre blood

The MAK-Collection Part IV: Biomonitoring Methods, Vol. 11.
DFG, Deutsche Forschungsgemeinschaft
Copyright © 2008 WILEY-VCH Verlag GmbH & Co. KGaA, Weinheim
ISBN: 978-3-527-31596-3

2,3,7,8-Tetrachlorodibenzo-p-dioxin

2,3,7,8-Tetrachlorodibenzo-p-dioxin (2,3,7,8-TCDD) may be formed as an undesirable by-product of thermal processes and of the incomplete combustion of organic substances in the presence of chlorine. It is also generated during the production of chlorophenols (e.g. 2,4,5-trichlorophenol) and phenoxyacetic acids.

The most relevant route of intake of 2,3,7,8-TCDD by the general population is with food. Moreover, intake via inhalation and dermal absorption are routes of relevance to workers [1]. After oral intake of 105 ng of [1,6-^{3}H]-2,3,7,8-TCDD by a 42-year-old man 87% was absorbed and 11.5% was excreted within 3 days. Thereafter only 0.03% of the dose was excreted daily until the 125th day [2]. The bioavailability is less than 1% following dermal exposure [3]. A penetration rate of 6 to 170 pg/cm^2/h through the human skin was determined *in vitro* [4]. The absorption by inhalation is estimated at 0.02 pg TEQ/kg/day [1]. 2,3,7,8-TCDD is mainly accumulated in the adipose tissue. Comparably high concentrations may also be present in blood fat [5]. Based on the lipid content, 2,3,7,8-TCDD concentrations that are distinctly lower that those measured in other organs are found only in the brain [1]. Metabolism of 2,3,7,8-TCDD produces polar metabolites (phenols), which are mainly excreted with the faeces after undergoing conjugation reactions in some cases. The mean elimination half-lives are about 7 years in man [5]. 2,3,7,8-TCDD can be excreted in mother's milk without being metabolised [5].

2,3,7,8-TCDD must be considered as a substance with carcinogenic potential for humans. However, no appreciable risk of cancer in man is to be expected, provided compliance with the MAK value (maximum permissible concentration at the workplace) of 10 pg/m^3 is ensured, 2,3,7,8-TCDD has therefore been assigned to Category 4 [6, 7]. This MAK value is not based on the effects to human health, but relates to the background exposure of the population (2.4 pg/g fat). It has been calculated that the 95% confidence limit of 4.8 pg/g blood fat will be reached if 10 pg/m^3 of 2,3,7,8-TCDD are inhaled for 8 hours per day. Furthermore, the Commission assigned 2,3,7,8-TCDD to Pregnancy Risk Group C. A comprehensive treatise on the toxicological aspects of 2,3,7,8-TCDD is presented in the MAK value documentation, and an overview of 2,3,7,8-TCDD exposure during production is also to be found there [6, 7].

In 1996 a mean 2,3,7,8-TCDD concentration of 2.4 pg/g fat (equivalent to 12 pg/L blood) was ascertained in the general population (n=180). The source of the background exposure of the population to 2,3,7,8-TCDD is assumed to be intake with food [5]. 2,3,7,8-TCDD concentrations of 828 and 56000 pg/g fat (equivalent to an intake of 10 µg/kg body weight) were found in the blood of residents of Seveso [8]. The 2,3,7,8-TCDD concentrations were determined from 1993 to 2000 in a group of

8 employees in a trichlorophenol production plant who were exposed to 2,3,7,8-TCDD. Recent measurements showed a mean 2,3,7,8-TCDD concentration of 226 pg/g fat (minimum value: 66 pg/g fat; maximum value: 480 pg/g fat) [9].

Authors: *J. Lewalter, G. Leng, W. Gries*
Examiner: *M. Ball*

2,3,7,8-Tetrachlorodibenzo-p-dioxin (TCDD)

Application Determination in blood

Analytical principle Capillary gas chromatography/ mass spectrometric detection (MS)

Contents

1 General principles
2 Equipment, chemicals and solutions
2.1 Equipment
2.2 Chemicals
2.3 Solutions
2.4 Calibration standards
3 Specimen collection and sample preparation
3.1 Specimen collection and storage
3.2 Preparation of the SPE cartridges
3.3 Sample preparation
4 Operational parameters
4.1 Operational parameters for gas chromatography
4.2 Operational parameters for mass spectrometry
5 Analytical determination
6 Calibration
7 Calculation of the analytical result
8 Standardisation and quality control
9 Evaluation of the method
9.1 Precision
9.2 Accuracy
9.3 Detection limit
9.4 Sources of error
10 Discussion of the method
11 References

The MAK-Collection Part IV: Biomonitoring Methods, Vol. 11.
DFG, Deutsche Forschungsgemeinschaft
Copyright © 2008 WILEY-VCH Verlag GmbH & Co. KGaA, Weinheim
ISBN: 978-3-527-31596-3

1 General principles

EDTA blood is initially subjected to liquid-liquid extraction. The analyte is subsequently enriched on an aluminium oxide solid phase and separated from matrix components. After separation by capillary gas chromatography, the analyte is measured by means of high-resolution mass spectrometry and electron impact ionisation (EI). Linear calibration graphs are used for quantitative evaluation. The standard solutions required for calibration are prepared with pooled human blood to which defined quantities of the calibration standard are added. The standard solutions are treated in the same manner as the blood samples to be investigated. ^{13}C-2,3,7,8-TCDD is added to the blood samples as an internal standard.

2 Equipment, chemicals and solutions

2.1 Equipment

Gas chromatograph with split/splitless injector, high-resolution mass spectrometer, autosampler and data processing system

Capillary gas chromatographic column:
Length: 30 m; inner diameter: 0.25 mm; stationary phase: 5% phenylmethylpolysiloxane; film thickness: 0.25 µm (e.g. DB 5-MSITD, J&W 122-4132)

Guard column:
Length: 5 m; Material: fused silica, deactivated; inner diameter: 0.53 mm (e.g. Agilent 160-2535-5)

Capillary connector (e.g. Restek 20429)

Disposable syringes containing an anticoagulant (e.g. 7 mL EDTA Vacutainer from Beckton-Dickinson)

Chromatography column made of glass:
Length: 14 cm; diameter: 1 cm (e.g. empty extraction vessels from Merck, Extrelut, No. 15371)

Glass Pasteur pipettes

Analytical balance

Laboratory centrifuge

25 mL Measuring cylinder

100 mL Glass beaker

200 mL and 500 mL Screw-capped jars with lids

30 mL Centrifuge vessels with Teflon-coated screw cap

10 mL Screw-capped glass test-tubes

Nitrogen evaporator (e.g. from Pierce)

10 µL and 100 µL Transferpettes (e.g. from Brand)

Drying cupboard (e.g. from Heraeus)

250 µL Microvials 8AC for the autosampler (e.g. from Chromacol)

1.5 mL Screw-capped glass vial

10 mL Volumetric flask

Mechanical shaker

Solid phase extraction station with Teflon stopcocks (e.g. from Macherey-Nagel)

2.2 Chemicals

n-Hexane SupraSolv® (e.g. Merck, No. 1.04371.2500)

Dichloromethane SupraSolv® (e.g. Merck, No. 1.06054.2500)

Toluene SupraSolv® (e.g. Merck, No. 1.08389.2500)

Sodium sulphate for organic trace analysis (e.g. Merck, No. 1.06639.0500)

GC cotton-wool (e.g. Shimadzu, No. 201-35311)

Alumina B Super I (ICN 04568)

Perfluorokerosene (e.g. Fluka, No. 77275)

Standard solution of 2,3,7,8-TCDD (Promochem NBS SRM 1614)

Standard solution of ^{13}C-2,3,7,8-TCDD (Promochem NBS SRM 1614)

Deionised water (e.g. Milli-Q water)

Helium 5.0 (e.g. from Linde Gas)

2.3 Solutions

See the instructions for using washing solutions 1 and 2 in Section 9.4.

Washing solution 1 (hexane/dichloromethane, 4:1 (v/v)):
400 mL n-hexane and 100 mL dichloromethane are measured using a measuring cylinder and then thoroughly mixed in a 500 mL screw-capped jar.

Washing solution 2 (hexane/dichloromethane, 98:2 (v/v)):
490 mL hexane and 10 mL dichloromethane are measured using a measuring cylinder and then thoroughly mixed in a 500 mL screw-capped jar.

Elution solution (hexane/dichloromethane, 1 : 1 (v/v)):
100 mL hexane and 100 mL dichloromethane are measured using a measuring cylinder and then thoroughly mixed in a 200 mL screw-capped jar.

The washing and elution solutions must be freshly prepared before each analytical series.

2.4 Calibration standards

Starting solution for the calibration standard:
The certified 2,3,7,8-TCDD solution from Promochem (67.8 ng/mL in iso-octane) serves as the starting solution.

Stock solution:
147 µL of the 2,3,7,8-TCDD solution (67.8 ng/mL) are pipetted into a 1.5 mL autosampler vial, which is then filled up with 853 µL toluene (10 ng/mL).

Working solution A:
100 µL of the stock solution are pipetted into a 1.5 mL autosampler vial, which is then filled up with 900 µL toluene (1 ng/mL).

Working solution B:
100 µL of working solution A are pipetted into a 1.5 mL autosampler vial, which is then filled up with 900 µL toluene (0.1 ng/mL).

Working solution C:
100 µL of working solution B are pipetted into a 1.5 mL autosampler vial, which is then filled up with 900 µL toluene (0.01 ng/mL).

Solution of the internal standard (ISTD)

The certified ^{13}C-2,3,7,8-TCDD solution from Promochem (65.9 ng/mL in iso-octane) serves as the starting solution.

759 µL of the ^{13}C-2,3,7,8-TCDD solution (65.9 ng/mL) are pipetted into a 10 mL volumetric flask and the flask is filled to its nominal volume with toluene (5 ng/mL).

The stock and working solutions are stored in sealed vials at $-18\,°C$, and they are stable under these conditions for at least 6 months.

The standard solutions used for calibration are prepared by spiking 10 mL of pooled blood in a 30 mL centrifuge vial in each case. The pipetting scheme for the preparation is shown in Table 1. The error resulting from dilution is negligible. These standard solutions must be freshly prepared before each analytical series.

Table 1. Pipetting scheme for the preparation of the calibration standard solutions in pooled blood

Volume of working solutions			Volume of the solution of the ISTD [µL]	Volume of blood [mL]	Concentration of the standard solution [pg/L]
A [µL]	B [µL]	C [µL]			
–	–	10	5	10	10
–	–	20	5	10	20
–	5	–	5	10	50
–	10	–	5	10	100
–	20	–	5	10	200
5	–	–	5	10	500
10	–	–	5	10	1000

3 Specimen collection and sample preparation

3.1 Specimen collection and storage

The blood withdrawn in EDTA glass Vacutainers is either analysed immediately or stored in the deep-freezer at $-20\,°C$. If stored in this way, the blood can be kept for at least 12 months. Depending on the size of the Vacutainers used, several tubes of blood may be necessary, as 10 mL of EDTA blood are required for the analysis.

3.2 Preparation of the SPE cartridges

A small quantity of GC cotton-wool is placed in a glass chromatography column and positioned at the outlet using a sharp object. Then the outlet is sealed with a Teflon stopcock and the column is mounted on a solid phase extraction station. Approx. 10 mL n-hexane are subsequently poured into the sealed chromatography column, which is then filled with 2 g Alumina B Super I. This filling is then overlaid with 1 g sodium sulphate.

3.3 Sample preparation

10 mL of the EDTA blood are transferred into a 30 mL centrifuge vial using a Transferpette. 10 mL deionised water and 5 µL of the ISTD solution are added and the sample is subsequently shaken thoroughly for 2 minutes. Then 10 mL hexane are added and the contents are mixed thoroughly on a mechanical shaker for 20 min. The mixture is then centrifuged at 2200 g for 10 min. If phase separation is incomplete, the sample is deep-frozen for approx. 2 hours. Then the mixture is thawed and the hexane phase is transferred using a Pasteur pipette to the Alumina B Super I glass column prepared as described in Section 3.2. After the Teflon stopcock has been opened, the supernatant hexane is allowed to drain at atmospheric pressure,

whereby the column should not run dry. Then the column is washed with 20 mL of washing solution 1 or 2 (see the relevant instructions in Section 9.4) and the eluate is discarded. The analyte is then eluted with 20 mL of the elution solution under a slight vacuum into two screw-capped test-tubes. The eluates are evaporated to approx. 3 mL in a stream of nitrogen, then they are combined and further evaporated to approx. 200 µL. This solution is transferred into a 250 µL microvial, evaporated to dryness with nitrogen, and finally taken up in 10 µL toluene. The microvial is tightly sealed. 2 µL of this solution are injected splitless into the GC-MS system.

4 Operational parameters

The analytical measurements are carried out on a GC-HRMS system with a split/splitless injector, autosampler and high-resolution mass spectrometer.

4.1 Operational parameters for gas chromatography

Capillary column:	Material:	Fused silica
	Stationary phase:	DB 5-MSITD
	Length:	30 m
	Inner diameter:	0.25 mm
	Film thickness:	0.25 µm
Guard column:	Material:	Fused silica
	Stationary phase:	without, deactivated
	Length:	1.2 m
	Inner diameter:	0.53 mm
	Connector:	Capillary connector
Temperatures:	Column:	1 minute at 80 °C; then increase at a rate of 20 °C/minute to 300 °C; 10 minutes at the final temperature
	Injector:	280 °C
	Transfer line:	250 °C
Carrier gas:	Helium (pre-pressure 80 kPa)	
Sample volume:	2 µL splitless	

4.2 Operational parameters for mass spectrometry

Ionisation type:	Electron impact ionisation (EI)
Electron energy:	43 eV

Ion source temperature: 250 °C

Accelerating voltage: 8000 V

Mulitplier: 400 V

Trap current: 0.5 mA

Resolution: 10 000

All other parameters must be optimised in accordance with the manufacturer's instructions.

5 Analytical determination

The operational parameters for the instruments are adjusted and 2 µL of the analytical sample are injected into the gas chromatograph in each case.
The masses used for quantification and the retention times of 2,3,7,8-TCDD and ^{13}C-2,3,7,8-TCDD (ISTD) are shown in Table 2.

Table 2. Retention times and recorded masses

Analyte	Retention time [minutes]	Recorded mass [m/z]
2,3,7,8-TCDD	10.48	319.8965
		321.8937*
^{13}C-2,3,7,8-TCDD (Internal standard)	10.50	331.9368
		333.9338*

The masses marked * are used for quantitative evaluation.

If the 2,3,7,8-TCDD content of a sample is above the linear range of the calibration function, then the blood sample must be appropriately diluted with deionised water before processing and then analysed anew.
In Figure 1 a chromatogram of a calibration standard solution (concentration 10 pg/L) in pooled blood is shown below (the same concentration in water is shown above, see Section 10).
Two quality control samples are analysed with each analytical series. In addition, a reagent blank is analysed in each analytical series. Deionised water is subjected to the processing described above instead of blood.

6 Calibration

The calibration standard solutions prepared as described in Section 2.4 are processed in the same manner as the samples (Section 3.3) and analysed by gas chromatography/mass spectrometry as stipulated in Sections 4 and 5. Calibration curves are obtained by plotting the quotients of the peak areas for 2,3,7,8-TCDD with that of the internal standard (^{13}C-2,3,7,8-TCDD) as a function of the concentrations used. The slope of the calibration function and the intercept with the y-axis are calculated by linear regression.

When 10 mL blood were used, a linear measurement range of 10 to 1000 pg/L was determined on the instrument described here. A new calibration curve is plotted for each analytical series.

Figure 2 gives an example of the linear calibration function for 2,3,7,8-TCDD in blood (and in water, see Section 10).

7 Calculation of the analytical result

The 2,3,7,8-TCDD concentration in blood samples is calculated on the basis of a linear calibration function (see Section 6). Quotients are calculated by dividing the peak areas of the analyte by that of the internal standard. These quotients are used to read off the pertinent concentration of 2,3,7,8-TCDD in pg per litre blood from the relevant calibration graph.

If the blood used to prepare the calibration standards exhibits background contamination with 2,3,7,8-TCDD, the resulting calibration graph must be shifted in parallel so that it passes through the zero point of the coordinates (the concentration of the background contamination can be read off from the intercept with the y-axis before the parallel shift). Any reagent blank values must be subtracted from the analytical results for the real samples. If the real samples have been previously diluted, the analytical result must be multiplied by the dilution factor.

8 Standardisation and quality control

Quality control of the analytical results is carried out as stipulated in the guidelines of the Bundesärztekammer (German Medical Association) [10] and in the special preliminary remarks to this series. Each analytical series should include two blood samples of known analyte content for the purpose of quality assurance. As material for quality control is not commercially available, it must be prepared in the laboratory. For this purpose defined quantities of 2,3,7,8-TCDD are added to pooled human blood (spiked with e.g. 30 pg/L). A six-month supply of this control material is prepared, divided into aliquots in sealable glass vessels and stored in the deep-freezer. The theoretical value and the tolerance range for this quality control material are determined in the course of a pre-analytical period (one analysis of the control material on 20 different days) [11, 12].

9 Evaluation of the method

9.1 Precision

The precision in the series was determined using human whole blood that was spiked with two different quantities of 2,3,7,8-TCDD, and then processed and analysed as described in the preceding sections. Six replicate assays of the blood samples yielded the precision in the series documented in Table 3.

Table 3. Precision in the series for the determination of 2,3,7,8-TCDD in blood (n=6)

Concentration [pg/L blood]	Standard deviation (rel.) [%]	Prognostic range [%]
10	9.1	22.3
30	10.4	24.4

Furthermore, the precision from day to day was checked using pooled human blood that contained approx. 100 pg/L 2,3,7,8-TCDD. The sample was processed and analysed on 6 different days. The resulting precision from day to day is given in Table 4.

Table 4. Precision from day to day for the determination of 2,3,7,8-TCDD in blood (n=6)

Concentration [pg/L blood]	Standard deviation (rel.) [%]	Prognostic range [%]
100	7.0	17.3

9.2 Accuracy

Recovery experiments were carried out at two different concentrations to test the accuracy of the method. The same material was used as that for the determination of the precision in the series. The blood samples were each processed and analysed 6 times as stipulated in the previous sections. The mean relative recovery rates are shown in Table 5.

Table 5. Mean, relative recovery rates for 2,3,7,8-TCDD in blood samples

n	Spiked concentration [pg/L blood]	Mean relative recovery [%]
6	30	106
6	100	108

Moreover, investigations were carried out to ascertain the losses of 2,3,7,8-TCDD that occurred during processing. For this purpose the spiked samples that were used to determine the precision in the series (blood 100 pg/L) and from day to day (pooled blood 100 pg/L) were also evaluated in direct comparison with pure standard solutions that were not subjected to processing. The mean absolute recovery rates thus obtained are presented in Table 6.

Table 6. Mean, absolute recovery rates for 2,3,7,8-TCDD in blood samples

n	Spiked concentration [pg/L blood]	Mean absolute recovery [%]
6	100	78
6	100	70

9.3 Detection limit

The detection limit for 2,3,7,8-TCDD was calculated as three times the signal/background noise ratio in the temporal vicinity of the analyte signal. The detection limit achieved by this procedure is 6 pg/L blood, the quantitation limit is 18 pg/L blood.

9.4 Sources of error

The activity of the Alumina B Super I that is used depends greatly on the quality and the freshness of the aluminium oxide. The solvent hexane/dichloromethane 4:1 (v:v) mixture described for the washing step functions only when the aluminium oxide is taken from a freshly opened original container. It is advisable to store the aluminium oxide for several days at 180 °C in the drying cupboard. The solvent mixture that is required in that case, consisting of hexane/dichloromethane 98:2 (v:v), permits reproducible elution and minimises the efforts required for activity testing. No plastic vessels were used in order to avoid adsorption. Broadening of the peak became gradually apparent after approx. 20 injections, which indicates occupation of the guard column. The problem can be eliminated by shortening or replacing the guard column.

10 Discussion of the method

The analytical method presented here is capable of detecting 2,3,7,8-TCDD in blood in the concentration range of relevance to occupational medicine. The very time-consuming and costly analytical method used in environmental medicine is suitable for detecting all the environmentally relevant PCDDs and PCDFs. However, this is not necessary for the range of interest to occupational medicine. The present analytical method exclusively determines highly toxic 2,3,7,8-TCDD of relevance to occupational medicine, and this is achieved with comparatively less effort and low material resources. Due to the selective separation of 2,3,7,8-TCDD from all the other interfering TCDD congeners on Alumina B Super I [13] a detection limit of 6 pg/L that is sufficient for the purposes of occupational medicine can be achieved by using only 10 mL of human blood. Arochlor mixtures were used to test if interference occurs due to PCBs. This could not be confirmed.

Concentrations are preferably expressed in pg/g of blood fat in environmental analysis of the dioxins. However, this requires an exact determination of the blood fat content. From the view of occupational medicine this relationship with the fat content is not necessary. Therefore the results for dioxin are based on the blood volume from which it was originally isolated. In general, the fat content of the blood is between 0.2 and 0.5%. Therefore 10 pg 2,3,7,8-TCDD/L blood is equivalent to about 2 to 5 pg 2,3,7,8-TCDD/g blood fat. When 40 mL of EDTA blood is used, a detection limit of 1 to 2.5 pg/L blood (0.5 pg/g blood fat) may be attained. However, this detection range is only necessary to detect the background exposure of the population.

The procedure was developed and validated using matrix calibration in pooled human whole blood. It is preferable to perform such a calibration in order to exclude matrix effects. However, the method was also tested with aqueous calibration (linear between 10 and 1000 pg/L). As calibration is performed with a labelled internal standard that can compensate for many analytical fluctuations, calibration can also be carried out in water if insufficient pooled blood is available.

Using this method it is possible to make a substantiated statement on the risk potential of intoxication with 2,3,7,8-TCDD within the period of a day. This is often important e.g. for victims and fire-fighters in the case of a fire.

Instruments used:
GC-HRMS system consisting of: HP 5890 II plus gas chromatograph with split/splitless injector, CTC A 200 S autosampler, and AutoSpec Ultima sector field MS.

11 References

[1] *D. Jung* and *J. Konietzko:* Polychlorierte Dibenzo-para-dioxine und Dibenzofurane, In: *J. Konietzko, H. Dupuis (eds.):* Handbuch der Arbeitsmedizin, 11th supplement. ecomed (1994).
[2] *H. Poiger* and *C. Schlatter:* Pharmacokinetics of 2,3,7,8-TCDD in man. Chemosphere 15, 1489–1494 (1986).
[3] *M. van den Berg, J. DeJongh, H. Poiger* and *J.R. Olsen:* The toxicokinetic and metabolism of polychlorinated dibenzo-p-dioxins (PCDDs) and dibenzofurans (PCDFs) and their relevance for toxicity. Crit. Rev. Toxicol. 24, 1–74 (1994).
[4] *L.W. Weber, K. Zesch* and *K. Rozman:* Penetration, distribution and kinetics of 2,3,7,8-tetrachlorodibenzo-p-dioxin in human skin in vitro. Arch. Toxicol. 65, 421–428 (1991).
[5] *World Health Organization:* Polychlorinated Dibenzo-para-dioxins and Dibenzofurans, Environmental Health Criteria, 88 (1989).
[6] *Deutsche Forschungsgemeinschaft:* List of MAK and BAT Values 2006, 42nd report, Wiley-VCH, Weinheim (2006).

[7] *H. Greim (ed.):* 2,3,7,8-Tetrachlordibenzo-p-dioxin. Gesundheitsschädliche Arbeitsstoffe. Toxikologisch-arbeitsmedizinische Begründungen von MAK-Werten. Issues 11, 19, 28, 34. Wiley-VCH, Weinheim (1986, 1993, 1999, 2002).

[8] *P. Mocarelli, L. L. Needham, A. Marocchi, D. H. Patterson, D. G. Brambilla, P. M. Gerthoux, L. Meazza* and *V. Carreri:* Serum concentrations of 2,3,7,8-tetrachlorodibenzo-p-dioxin and test results from selected residents of Seveso, Italy. J. Toxicol. Environ. Health 32, 357–366 (1991).

[9] *J. Lewalter* and *W. Gries:* Unpublished BM findings of follow-up investigations on employees formerly involved in trichlorophenol production (1998).

[10] *Bundesärztekammer:* Richtlinie der Bundesärztekammer zur Qualitätssicherung quantitativer laboratoriumsmedizinischer Untersuchungen. Dt. Ärztebl. 100, A3335 – A3338 (2003).

[11] *J. Angerer* and *G. Lehnert:* Anforderungen an arbeitsmedizinisch-toxikologische Analysen – Stand der Technik. Dt. Ärztebl. 37, C1753–C1760 (1997).

[12] *G. Lehnert, J. Angerer* and *K. H. Schaller:* Statusbericht über die externe Qualitätssicherung arbeits- und umweltmedizinisch-toxikologischer Analysen in biologischen Materialien. Arbeitsmed. Sozialmed. Umweltmed. 33(1), 21–26 (1998).

[13] *H. Hagenmaier, H. Brunner, R. Haag, H.-J. Kunzendorf, M. Kraft, K. Tichaczek* and *U. Weberruß:* Stand der Dioxin Analytik VDI Report No.: 604 (1987).

Authors: *J. Lewalter, G. Leng, W. Gries*
Examiner: *M. Ball*

2,3,7,8-Tetrachlorodibenzo-p-dioxin

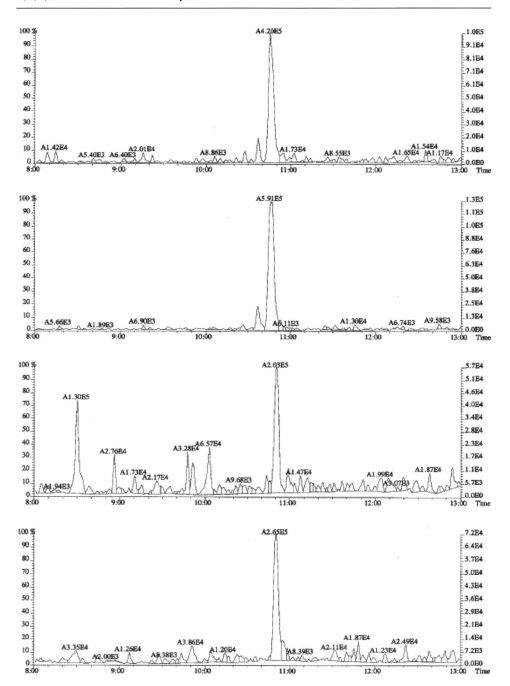

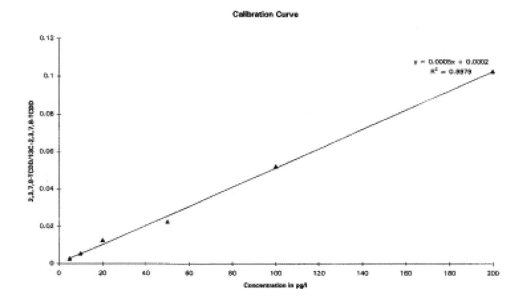

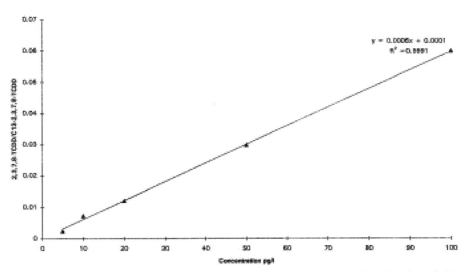

Fig. 2. Linear calibration function of 2,3,7,8-TCDD, prepared in water (above) and pooled human blood (below)

◀

Fig. 1. GC/MS-SIM chromatogram of a processed calibration standard solution in water (above) and pooled blood (below), concentration 10 pg/L in each case. Detected mass fragments: in each case m/z = 319.8965 and 321.8937

Members, Guests and ad hoc Experts of the Working Subgroup

Analyses of Hazardous Substances in Biological Materials of the Commission of the Deutsche Forschungsgemeinschaft for the Investigation of Health Hazards of Chemical Compounds in the Work Area

Leader: Prof. Dr. J. Angerer
Institut für Arbeits-, Sozial- und Umweltmedizin
Universität Erlangen-Nürnberg
Schillerstraße 25/29
D-91054 Erlangen

Deputy leader: Dipl.-Ing. K.-H. Schaller
Institut für Arbeits-, Sozial- und Umweltmedizin
Universität Erlangen-Nürnberg
Schillerstraße 25/29
D-91054 Erlangen

Members: PD Dr. M. Bader
Institut und Poliklinik für Arbeitsmedizin
der Medizinischen Hochschule Hannover
Carl-Neuberg-Straße 1
D-30625 Hannover

Dr. M. Blaszkewicz
Institut für Arbeitsphysiologie an der
Universität Dortmund
Ardeystraße 67
D-44139 Dortmund

PD Dr. T. Göen
Institut für Arbeits-, Sozial- und Umweltmedizin
Universität Erlangen-Nürnberg
Schillerstraße 25/29
D-91054 Erlangen

Prof. Dr. G. Leng
BAYER Industry Services GmbH & Co.OHG
SUA-Gesundheitsschutz-Biomonitoring
Gebäude L 9
D-51368 Leverkusen

PD Dr. M. Müller
Georg-August-Universität Göttingen
Abteilung für Arbeits- und Sozialmedizin
Waldweg 37
D-37073 Göttingen

Guests:
Dr. J. Begerow
Hygiene-Institut des Ruhrgebiets
Institut für Umwelthygiene und Umweltmedizin
Rotthauser Straße 10
D-45879 Gelsenkirchen

Prof. Dr. W. Butte
Carl-von-Ossietzky-Universität Oldenburg
FB Chemie
Postfach 25 03
D-26129 Oldenburg

Prof. Dr. L. Dunemann
Hygiene-Institut des Ruhrgebiets
Institut für Umwelthygiene und Umweltmedizin
Rotthauser Straße 10
D-45879 Gelsenkirchen

Prof. Dr. H. Emons
European Commission – Joint Research Centre
Institute for Reference Materials and Measurements
Retieseweg 111
B-2440 Geel
Belgium

Dr. H.-W. Hoppe
Medizinisches Labor Bremen
Dr. Wittke, Dr. Gerritzen und Partner
Haferwende 12
D-28357 Bremen

Dr. H.-U. Käfferlein
Berufsgenossenschaftliches Forschungsinstitut für Arbeitsmedizin
Bürkle-de-la-Camp-Platz 1
D-44789 Bochum

Dr. H. M. Koch
Berufsgenossenschaftliches Forschungsinstitut für Arbeitsmedizin
Bürkle-de-la-Camp-Platz 1
D-44789 Bochum

Dr. B. Michalke
GSF-Forschungszentrum f. Umwelt u. Gesundheit GmbH
Ingolstädter Landstraße 1
D-85764 Neuherberg

Dr. G. Müller
Institut für Hygiene und Arbeitsmedizin
der Universität GH Essen 1
Hufelandstraße 55
D-45147 Essen

Prof. Dr. E. Richter
Walther-Straub-Institut für Pharmakologie und Toxikologie
Goethestraße 33, Raum 214
D-80336 München

Prof. Dr. G. Sabbioni
Tulane University School of Public Health and Tropical Medicine
Department of Environmental Health Sciences
1440 Canal St., Suite 2100
New Orleans, LA 70112-2704
USA

PD Dr. G. Scherer
ABF Analytisch-Biologisches Forschungslabor München
Goethestraße 20
D-80336 München

Prof. Dr. P. Schramel
GSF-Forschungszentrum f. Umwelt u. Gesundheit GmbH
Ingolstädter Landstraße 1
D-85764 Neuherberg

Members, Guests and ad hoc Experts of the Working Subgroup

PD Dr. W. Völkel
Bayerisches Landesamt für Gesundheit und Lebensmittelsicherheit
(LGL Bayern)
Pfarrstr. 3
D-80538 München

Dr. W. Will
BASF AG
GOA/CB – H 306
D-67056 Ludwigshafen

ad hoc experts: Dr. D. Barr
Centers for Disease Control
Dept. of Health & Human Service
4770 Buford Highway, N.E. Atlanta, GA 30347-3724
USA

Dr. E. Berger-Preiß
Fraunhofer Institut für Toxikologie
und Experimentelle Medizin (ITEM)
Arbeitsgruppe: Analytische Chemie
Nikolai-Fuchs-Straße 1
D-30625 Hannover

Dr. J. Cocker
Health & Safety Laboratory
Harpur Hill, Buxton SK17 9JN
UK

Dr. J. Hardt
Institut für Laboratoriumsmedizin, Mikrobiologie
und Umwelthygiene Klinikum Augsburg
Stenglinstr. 2
D-86156 Augsburg

Dr. P. Heitland
Medizinisches Labor Bremen
Dr. Wittke, Dr. Gerritzen und Partner
Haferwende 12
D-28357 Bremen

Prof. Dr. O. Herbarth
Umweltmedizin & Umwelthygiene
Medizinische Fakultät
Universität Leipzig
Liebigstr. 27
D-04103 Leipzig

Dr. S. Pavanello
Università Degli Studi Di Padova
Dipartimento Di Medicina Ambientale e Sanità Pubblica
Via Giustiniani 2
35128 Padova
Italia

Dr. B. Rossbach
Institut für Arbeits-, Sozial- und Umweltmedizin
Obere Zahlbacher Str. 67
D-55131 Mainz

Dr. T. Schettgen
Institut für Arbeits- und Sozialmedizin der RWTH Aachen
Pauwelsstraße 30
D-52074 Aachen

Dr. P. Simon
Institut national de Rechere et Sécurité (INRS)
Service SBSO (Surveillance Biologique aux Substances Organiques), Département PS (Polluants et Santé)
Avenue de Bourgogne, BP 27
54501 Vandoeuvre Cedex
France

Dr. T. Weiss
Berufsgenossenschaftliches Forschungsinstitut für Arbeitsmedizin
Bürkle-de-la-Camp Platz 1
D-44789 Bochum

Scientific secretariat:
O. Midasch
Institut für Arbeits-, Sozial- und Umweltmedizin
Universität Erlangen-Nürnberg
Schillerstraße 25/29
D-91054 Erlangen